Studies in the Philosophy of Sociality

Volume 9

More information about this series at http://www.springer.com/series/10961

Raul Hakli • Johanna Seibt
Editors

Sociality and Normativity for Robots

Philosophical Inquiries into Human-Robot Interactions

Springer

Editors
Raul Hakli
Department of Political and Economic Studies
University of Helsinki
Helsinki, Finland

Johanna Seibt
Research Unit for Robophilosophy
Department for Philosophy and History of Ideas
School of Culture and Society
Aarhus University, Aarhus, Denmark

Studies in the Philosophy of Sociality
ISBN 978-3-319-53131-1 ISBN 978-3-319-53133-5 (eBook)
DOI 10.1007/978-3-319-53133-5

Library of Congress Control Number: 2017938836

Printed on acid-free paper

This Springer imprint is published by Springer Nature
The registered company is Springer International Publishing AG
The registered company address is: Gewerbestrasse 11, 6330 Cham, Switzerland

Preface

Social robots, if used pervasively in society, will change the fabric of human social interactions more profoundly than any other technology before – or so we have currently reason to believe. Since human social interactions realize human values, any change in social interactions also potentially affects the network of our sociocultural and ethical values. That social robotics presents us with far-reaching ethical questions has been observed at least a decade ago. Since this time, however, it also has become clear that the way in which humans react to social robots does not allow for normative assessments using common subsumptive methods of applied ethics. The empirical facts of human interactions with social robots show that these interactions cannot be conceived in familiar categories of human-machine interaction or human-computer interaction. Rather, it appears that we need to explore whether the notions of sociality and normativity, the hallmarks of human-human interactions, can be suitably extended to capture the phenomena of human interactions with so-called "social" robots. Only in tandem with addressing the descriptive tasks of social robotics can adequate ethical assessments be formulated.

Thus social robotics presents us with a formidable challenge in descriptive as well as in ethical regards. In answer to this challenge, a new area of philosophy has constituted itself which aims to come to terms with the very idea of artificial social agency – "robophilosophy." The articles we have collected here – some of which have grown out of earlier articles presented at the inaugural 2014 event of the Robo-Philosophy series (Seibt, J., Hakli, R., and Nørskov, M., eds.: *Sociable Robots and the Future of Social Relations: Proceedings of Robo-Philosophy 2014*, IOS Press) – for the first time take up the narrowly conceptual-analytical tasks introduced by *social* robotics. While human-computer interaction has been investigated with broad interdisciplinary scope including philosophy, the interdisciplinary field of HRI (human-robot interaction) largely still proceeds without the contributions of humanities research in general and of philosophy in particular.

The main aim of this book is to demonstrate that robo-ethics is not enough – we need to engage all philosophical disciplines, in particular also theoretical disciplines such as ontology and philosophy of mind, in a joint interdisciplinary endeavour of coming to terms with the descriptive and normative tasks social robotics is putting before us. Even though the contributions to this book address the research community in philosophy, they are accessible to an interdisciplinary readership; thus the book also lends itself as textbook for graduate courses on human-robot interaction taught from a broad range of different disciplinary angles.

We thank all those who made the book possible. Most of all, we would like to thank the authors of this volume for their contributions and their patience with the production process. But we also wish to thank Marco Nørskov, co-organizer of the Robo-Philosophy 2014 conference, who has been promoting philosophy of social robotics at the Aarhus Philosophy Department since 2008; it is ultimately due to him that we, jointly with other philosophers in Aarhus, began to explore the rich systematic perspectives of this field. We would like to thank all members of the "Research Group for Philosophical Enquiries in Social Robotics" (PENSOR group, 2012–2015) and its successor, the "Research Unit for Robophilosophy," for many fruitful discussions: Raffaele Rodogno, Jens-Christian Bjerring, Jacob Busch, Ryuji Yamazaki, Stefan Larsen, Malene Flensborg Damholdt, Christina Vestergård, Glenda Hannibal, Thea Puggaard Strand, Rikke Mayland Olsen, and Søren Schack Andersen. We are grateful to Arto Laitinen for reading and commenting on the first draft, and we thank the series editor Raimo Tuomela for admitting into the series a new perspective on the current and future significance of social ontology research. Finally, we gratefully acknowledge the support of the Velux Foundation for the research project "Social Robotics and Human Self-Understanding" (2012–2015), in the context of which this book was produced.

Helsinki, Finland
Aarhus, Denmark
September 2016

Raul Hakli
Johanna Seibt

Contents

Chapter 1
"Sociality and Normativity for Robots": An Introduction

Raul Hakli and Johanna Seibt

Official projections predict that in the course of the next two decades human societies will see pervasive use of robotic technology in all contexts of social interaction, public and private. Currently, so-called "social robots" are designed for use in elderly care and education, with applications ranging from cognitive training to language tutoring to autism therapy and research. Social robots are also used as "tour guides", "diet coaches", "personal assistants", and "cleaners", and it is predicted that by 2020 they will "influence every aspect of work and home" (euRobotics aisbl, 2013).

The accuracy of these predictions remains to be seen, but it is safe to say that there will be changes that can profoundly affect our cultural practices and social relationships. Of course, technological change has taken place throughout the history of human kind, and in people's experience the rate of change is often seen as increasing and its effects as more dramatic than before. During the past three decades we have experienced in tight sequence how the introduction of new technology has changed our everyday lives. The first wave of change occurred when personal computers entered our workplaces and our homes; the second and even farther reaching change in our social practices set in when mobile phones and then smart phones delocalized communication and information, making us accessible nearly everywhere. The ubiquitous modifications of "information technology (IT)" in the first phase were compounded by the impact of what we came to label

R. Hakli (✉)
Department of Political and Economic Studies, University of Helsinki, P.O. Box 24 (Unioninkatu 40A), FI-00014, Helsinki, Finland
e-mail: raul.hakli@helsinki.fi

J. Seibt
Research Unit for Robophilosophy, Department for Philosophy and History of Ideas, School of Culture and Society, Aarhus University, Denmark Jens Chr. Skous Vej 7, DK-8000, Aarhus C, Denmark
e-mail: filseibt@cas.au.dk

R. Hakli, J. Seibt (eds.), *Sociality and Normativity for Robots*, Studies in the Philosophy of Sociality 9, DOI 10.1007/978-3-319-53133-5_1

"information and communication technology (ICT)" in the second phase. Now we are on the brink of possibly even more "disruptive" reconfigurations of socio-cultural space due to what could be called the "information, communication, and robotic technology (ICRT)" on the horizon. Arguably, the proliferation of the use of robotic technology will have even greater effects than the previous two phases since robotic "social" agents put the new powers of information and communication technology into physical space, and thereby not only extend our social interaction space, but also structure and delimit it in novel ways.

These prospects call for philosophical reflections at two levels. On the one hand, there are ethical questions concerning specific robotics applications, especially if these replace human interaction partners. Until now concerns in the area of applied ethics ("roboethics") have dominated philosophical discussions involving social robotics. On the other hand, social robotics raises also fundamental conceptual and metaphilosophical issues that philosophers need to address in addition to or even before answering ethical questions about specific applications. Social robotics technology is a far cry from the mechanization of our physical environment that we know from vending machines and automated cash counters in supermarkets – in fact, as has been noted variously in the new area of "Human-Robot Interaction" (HRI) research, we do not yet possess the right categories to classify this new sort of social agent and our interactions with it.

There is something deeply unsettling about the concept of a social robot, both unnerving and enticing in its possibilities. Social robotics is not only the engineering of robotic movements, it is the engineering of human social actions. So far we have been shaping the actions of our fellow human beings by means of ideas, rules, norms, incentives, physical objects, and environments, but none of what was human made could engender a social interaction – the only sort of item that could elicit a social response was another human agent, something that belonged to the sphere of sociality but was not manufactured. The unsettling aspect of social robotics lies in the fact that for the first time humans manufacture the kind of item that affords social interactions, or so it appears.

So-called "social" robots are designed to engage us in social interactions and to enable us to develop social relationships with them, and as empirical research shows, humans are quite willing to accept them as social partners. This is in conflict with our traditional conceptions of sociality and social interaction, which presuppose the essential capacities of human subjectivity (consciousness, rationality, intentionality, free will, normative agency etc.) in all partners of a social interaction. In short, the phenomena of human interactions with "social" robots contradict the longstanding notion that sociality is grounded in reciprocity. Since reciprocity-based conceptions of sociality are at the centre of the Western understanding of human nature, of our moral standing, our political authority, and our human rights, the very idea of artificial social agents, together with its empirical validations, presents a formidable challenge to Western philosophy.

In answer to this challenge the new area of philosophy is constituting itself which aims to come to terms with the very idea of artificial social agency – "robophilosophy" (Seibt, Hakli, & Nørskov, 2014). Robophilosophy is in most cases an interdisciplinary research effort in the intersection of philosophy, robotics, computer science, cognitive science, anthropology, psychology, and sociology. Robophilosophy is philosophy of, for, and by social robotics – it examines the socio-cultural implications of inserting a new type of agents into the space of human social interaction; it develops new conceptual tools that will help us to understand human-robot interactions and program robots in accordance with these; and it is experimental philosophy, using robots as a new investigative tool to explore the conditions of human social interaction (Seibt, Hakli, & Nørskov, 2018; Seibt, 2017).

This volume is a contribution to robophilosophy that builds a particularly important "access road" into this new field. By now several explorations into the conceptual and cultural implications of social robotics from a largely "continental" perspective in philosophy (Coeckelbergh, 2012; Gunkel, 2012; Nørskov, 2015) and "roboethics" and "machine ethics" have already staked their claims (e.g. Sullins, 2005, 2006; Sharkey & Sharkey, 2012; Wallach & Allen, 2009), researchers in analytical philosophy, and especially in social ontology, have yet to find their way into philosophy of social robotics. The present volume aims to create this important link, and to show that it is to mutual benefit.

The articles we have collected here for the first time take up the narrowly conceptual-analytical tasks introduced by social robotics. Thus robophilosophy as it is undertaken here is not so much reflective philosophy *of* social robotics, but rather constructive philosophy (and cognitive science) *for* social robotics. Keeping the analytical focus on sociality and normativity the chapters explore conceptual core questions arising with the very idea of an artificial social agent, such as the following: What does sociality mean in the context of social robotics? Can robots be social in the same sense as human beings are social? In which sense and to what extent will robots be able to simulate human sociality? In which sense of normative agency will robots be able to enter into normative relationships with human beings or with each other? What is the place of robotic entities in social ontology? Can robots be persons or normative agents? Can they become members of human societies?

To clarify these questions, one cannot resort to an "established and comprehensive" theory of human sociality, let alone of human-robot interactions, for there are none yet – interdisciplinary research on human sociality and human-robot interaction (HRI) both are young fields. Moreover, both research areas, though the second more than the first, struggle with the fact that we lack the theoretical vocabulary to articulate forms of sociality which deviate from the philosophical model case involving individual subjects with reciprocal capacities of intentionality and normative reasoning.[1]

[1] In a similar vein, the contributions to (Misselhorn, 2015) explore whether and how philosophical notions of collective agency and cooperation can be brought to bear on human interactions with artificial agents of all kinds. Since human interactions with virtual agents on-screen differ profoundly from human interactions with robots, i.e., artificial agency within the physical space of human social interaction, it is important, we believe, to differentiate the discussion. Moreover, we believe it useful to put the investigative focus first on the fact that the phenomena of human-robot

Chapter 2 thus introduces in outline a conceptual framework for the description of "asymmetric" – i.e., not fully reciprocal – forms of sociality. The framework for "Simulated Social Interactions" (SISI) which Johanna Seibt sketches here, can be used to formulate claims and investigatory hypotheses for empirical and conceptual studies of social interactions in general, and human-robot interactions in particular. She first draws attention to what she calls the "soft problem in the ontology of social robotics" – i.e., the task of describing human social interaction with robots without ascribing to robots beliefs and intentions from the outset. To address this problem, she argues, we need to treat the "as if" mentioned in many descriptions of human-robot interaction as the "as if" of simulation, not of fictional pretend-play. She distinguishes five kinds of simulation or partial realization, formally defined in terms of relationships between process systems. Then she shows how these five modes of simulation can be used to define for any type of two-agent interaction a matrix of "simulatory expansions" of that interaction type. Simulatory expansions of social interactions can be used to articulate different kinds and degrees of sociality. This allows us to clarify which conceptual and empirical reasons we traditionally draw on, and which we could draw on, when we determine the requirements for social interactions. It also allows us to identify precisely where, and on what grounds, we require different capacities and affordances in our interaction partners when it comes to social interaction with robots as opposed to, e.g., interactions with children or animals. In this way, she suggests, we can gradually, in a feedback loop of conceptual and empirical research, build up a taxonomy of social interactions, symmetric and asymmetric, for humans, robots, and other candidates for social interaction partners.

The subsequent chapters undertake this sort of research, combining terminological differentiations with substantive claims about sociality and normativity that are grounded in empirical research in cognitive science and psychology.

Chapter 3 contests the presuppositions of the traditional debate about sociality by a substantive account of sociality developed within a broader systematic context of a naturalist process metaphysics. Mark Bickhard, a protagonist of the "interactivist" paradigm in cognitive science and philosophy of cognition for many years, here shows how the basic questions about robot sociality are reconfigured once we proceed from the assumption that the world consists of processes rather than substances or other "static" types of entities. In a first step he recapitulates in broad strokes some central arguments of his earlier work, stressing that process ontology is in close contact with science at all scales and allows for a coherent account of emergence as dynamic organization of other processes. Some emergent processes are the dynamic organizations of systems far from thermodynamic equilibrium, i.e., self-maintaining and recursively self-maintaining systems, which, Bickhard argues, realize normative functionality in nature. On this thermodynamic account of normative function we can make sense of the idea that one process functionally presupposes another,

interaction challenge the classificatory concepts of social ontology at its most basic level, namely, the very notions of sociality and normativity themselves.

which, Bickhard argues, in turn can be used to develop an account of representation that is not troubled by "grounding" problems. In a second step, Bickhard sketches how elements of social reality (including language and personhood) arise in this naturalist framework. In order to solve coordination problems agents introduce "situation conventions," which are "emergent and implicit," that is, not based on any preceding reasoning. In a third step, Bickhard considers the possibility of robot sociality. The handling of situation conventions requires that the agent "has a stake in the world," i.e., interests that drive the establishment and repair of social relations. But "stakes," Bickhard claims, are tied to the architecture of biological far-from-equilibrium systems. Thus, Bickhard concludes, robots would be able to *be* social rather than merely simulate social behaviour if they were far-from-equilibrium systems with an architecture that enables emergent normativity akin to biological systems. While this may be possible, contemporary robots are not genuine social agents.

The chapter by David Eck and Alex Levine continues the theme of questioning longstanding assumptions in our understanding of sociality from the process-ontological perspective of the interactivist approach. Here the focus is on the traditional assumption that sociality presupposes individual subjectivity; from this assumption it would follow that robots, which are not individual subjects, cannot possibly engage in social relations. The authors begin by highlighting several paradoxical consequences of this assumption, which they call the "priority of individuality" thesis, in other areas of philosophy that share the following basic structural problem arising for traditional theories of sociality. As long as the explanandum, e.g., sociality, of the theory is conceived of as a relation, we need to postulate that there are different individuals between which such a relation can hold; but the relation in question is defined in such a way that the differences between these individuals are irrelevant. The structural problem of co-constitution of relatum and relation, or specifically, of individual and society – well familiar from the philosophy of German idealism – is best solved, the authors suggest, by turning to a purely process-based account, and specifically to the enactivist approach to cognition and sociality. With focus on the work by H. De Jaegher, the authors discuss recent proposals for which interactions must be added to largely autopoietic or self-maintaining processes to capture the characteristics of social coordination. Eck and Levine argue that these current process-based characterizations of sociality obscure, or even fail to accommodate, first-person experience and the historical dimension of subjectivity. To avoid that the process-based approach, which as such is on the right track, results in "hollow collectivism" it is important, the authors suggest in conclusion, to integrate data from qualitative research into the set of empirical data to be modelled.

The chapter by Jedediah Allen and Hande Ilgaz contributes an important piece of empirical research to the vision of a process-based theory of sociality suggested in the previous three chapters. The generic references to "social coordination" and "social relations" in Chaps. 3 and 4 raise the question of whether a process-based approach is strong enough also to describe the details of social learning, which in turn can be used to explain the complexity of social cognition and social behaviour in adults. Only if a purely process-based description of social learning is possible, one might argue, can one at all entertain the idea that the complex competences of sociality are not tied to a special kind of substance but could be implemented in a robotic system with the same process architecture. Allen and Ilgaz address this question by illustrating how one might explain social learning in general and imitation learning in particular from an interactivist or "action-based" perspective. They argue that imitation itself undergoes learning and development – selective imitation and over-imitation does not need to be understood as reflecting unique capacities of the human mind but can be treated as an instance of social meta-learning, i.e., of learning how to respond to the interaction possibilities of a learning situation that involves other agents. Such meta-learning capabilities will be required of social robots as well. On the interactivist account, an object is represented by its affordances for interaction; during imitation learning these affordance structures are modified but, this is the authors' decisive point, not only by the interaction with the object but by interaction with the entire situation in which the object is embedded and together with the affordances provided by other agents in the situation. In an empirical study Allen and Ilgaz show how this perspective on social meta-learning can be applied to an ongoing debate about why children imitate causally unnecessary actions while learning about a new artifact (i.e., over-imitate). The authors suggest that over-imitation in children shows that the affordance structures representing objects are acquired together with the social affordances of the situation and thus as "cultural artifacts."

The account of social cognition presented in the previous chapter does not presuppose the capacity of social agents to read each other's minds. The next chapter, by Vìctor Fernández Castro, endorses this view, this time not from the perspective of developmental psychology but from the discussion of social cognition in philosophy of mind, drawing on evolutionary biology. Fernández Castro first rehearses the traditional position of the "mindreading" approach, which presupposes that social agents can represent the other's mental states (beliefs, intentions, and emotions) and that this capacity is central for social behaviour that involves the prediction and explanation of the actions of others. In contrast, the "mindshaping" approach – developed by Mameli – holds that we can predict and explain the actions of others merely by assuming that others follow rational norms just as we do. Fernández Castro argues that the mindshaping approach is the superior model for human social cognition since – among a list of other strong arguments – unlike the mindreading approach it can explain both cases of cooperation without prediction and cases of prediction of actions. In a further step Fernández Castro identifies three areas where the mindshaping approach can be of use for social robotics. Due to the emphasis on developmental aspects mindshaping provides a testable model for "developmental robotics," which combines developmental psychology and social robotics research by implementing developmental models of social cognition. In addition, it can shed light on the relationship between general and social intelligence. Finally, mindshaping mechanisms may have direct implications for the design of companion robotics.

The next chapter can be read as an oblique demonstration of this last claim that the "mindshaping" approach can be used as a basic principle of software architecture. Here Felix Lindner and Carola Eschenbach develop a framework of affordances and affordance spaces that could be implemented and applied in order to enable social robots to spatially coordinate their actions in the presence of human beings and other robots. Nowhere in this framework are minds read or represented – the social behaviour is engendered by affordance spaces that reflect social norms. The authors note that although there have been several studies on affordances in robotics, most of them have focused on specific interaction contexts and paid too little attention to the effects that agents' behaviours can have on each other's affordances. Their focus is on spatial activities: They develop a framework in which various regions surrounding agents, objects, and, more generally, affordances can be defined in order for robots to be able to infer minimally obtrusive movement trajectories and locations to position themselves or other objects. One example would be avoiding overlaps between affordance spaces because they might indicate that action opportunities of other agents might become blocked. A careful analysis of affordance spaces during planning, for instance, enables robots to display behaviour that is more readily perceived as socially acceptable as the authors demonstrate via several examples and case studies.

The next three chapters investigate specific conceptual issues for human-robot interaction: Can humans and robot perform a joint action? Can they have joint commitment? And finally, which, if any, "understanding" of norms can humans and robots have in common?

In their contribution Aurélie Clodic, Elisabeth Pacherie, Rachid Alami, and Raja Chatila study the general conditions for joint action between humans and robots. They note that human joint action has been intensively studied within psychology and philosophy, and try to see to what extent the results of those studies can be connected to the field of human-robot interaction. The authors find parallels, in the single-agent case, between philosophical accounts of intentional action and layered architectures proposed in artificial intelligence and robotics. Such connections motivate them to look for similarities also in the case of joint action. They analyse several requirements and coordination processes that have been discussed in the case of human joint action and translate them to the case of human-robot joint action. Their study culminates in taking the first steps toward what they call a "framework for joint action" that describes three layers at which human-robot joint action can be represented. Such a framework can be used not only for analysing information needs and representational processes involved in joint action but potentially also for implementing robots capable of joint action with humans.

Alessandro Salice and John Michael focus on the notion of commitment which according to them is central in human social interaction including joint action. They will study the possibility of robots making commitments to humans and humans being motivated to honour commitments made to robots. They argue that even though full-blown commitments might be impossible in interactions between humans and robots, it might be useful to aim at designing robots that stimulate in humans tendency to act as if commitments were effective. This might suffice to acquire the potential social benefits that commitments generate such as predictability of action and higher level of engagement. Salice and Michael present several challenges that designing such robots might involve and consider ways to address them. In addition to studying ordinary interpersonal commitments they devote special attention to joint commitments that are associated with joint action, and consider whether humans could form groups with robots and whether psychological processes of group-identification could be triggered in cases of such mixed groups.

Johannes Brandl and Frank Esken investigate in which sense we can justifiedly say that robots are "social" agents, which they take to turn on the question in which sense we can justifiedly say that robots "understand norms." In interaction with empirical research Brandl and Esken develop three criteria which in combination characterize full-scale normative competence. Each one of these criteria defines a "stage" of social behaviour or "social intelligence," they suggest. At the first stage, agents conform to regularities that have a "social foundation" in the sense that they are behaviours of a group that are enforced by the dominant members of the group. This form of social behaviour, which is motivated by the individual's instrumental rationality, is also displayed by animals, e.g., by "rules" for the treatment of infants among chimpanzees. The second stage of social behaviour is reached when agents follow a rule because others expect it from them. When young children at the age of three protest against a violation of a norm they exhibit this level of understanding of norms or rules, Brandl and Esken argue. Here agents not only behave in accordance with a rule for instrumental reasons but they follow the rule *as* a rule, as a matter of "social practical rationality." The normative force of the rule here rests on social expectations. Finally, at stage three agents follow a rule with the understanding that the rule applies in all cases of this kind, i.e., the normative force of the rule lies in the rule itself. Only social behaviour at this last stage, Brandl and Esken suggest, involves a genuine understanding of norms as norms, and only at this last stage we can justifiedly speak of social agency in the full sense. To the extent that robots are unlikely to be able to operate on the third stage of social intelligence, we should refrain from considering them as social agents, even though they may exhibit social behaviour at the first two stages.

The final two chapters again broaden the scope of inquiry and consider the questions what we will need to become when we begin to share our interaction space with robots, how we will need to invent new venues to preserve our capacity to care, i.e., to act freely for the sake of values, and what robots would need to become if we were to integrate them into our life-form.

In the penultimate chapter, Antonio Carnevale approaches the question whether robots can figure as social or normative agents from the perspective of foundational reflections engaged by recent philosophy of technology. Modern electronic technology uses software and thus representations at levels where older technology merely employs causal translations; thus modern technology introduces new levels of abstractions or declarative representations. With care robots the abstractive declarations of technology enter into the most intimate region of human interaction space. These dimensions we need to have in mind, Carnevale argues, when we ask what kind of care relationship could exist between a robot and a human. The rational response, he suggests, is to reform certain concepts for social relations in such a way that interactions with technology are included from the outset. As long as we merely insist that care must be based on love or solidarity, we have only the option of deploring that the increasing use of "carebots" in the healthcare sector will decrease the "humanity" in caretaking. But we cannot give up caring, which is, in Carnevale's view, an essential aspect of a human life. Caring means setting values and choosing freely to pursue them. The dilemma: "no care in healthcare versus no technology in healthcare" can be avoided, Carnevale argues, once we recognize that there are other ways to care – we care when we are responsible. Thus the responsible integration of technology into healthcare and other social interaction spaces can become our new form of caring. Carnevale sketches three components that characterize such a responsible way of policy making that he calls the "I Tech Care" approach.

In the last chapter, Hans Bernhard Schmid considers the possibility of robots' participating in our "life-form" (i.e., our lifeform as macrobic organisms and our Wittgensteinian form of life as complex of social practices) and of them developing a life-form of their own. The chapter's main aim is to critique the notion of a robot by an analysis of several various misplaced dichotomies in the characterization of robots which, he suggests, hamper the current discussion about robots in society. Schmid begins with a closer look at the specific place where the term "robot" enters human cultural history, Karel Capek's play "Rossum's Universal Robots." Using an interpretation of Capek's play as an expository foil throughout, Schmid first argues against the distinction between routine work and discursive practices that underlies many "robo-sceptical" views. He draws on the distinction between "mimeomorphic" and "polymorphic" actions, i.e., actions with implicit and explicit complexity, and argues that it is by no means clear that robots cannot perform actions that simulate actions with implicit complexity, as these are involved in our social and discursive practices. The second dichotomy he questions is the division between natural life and mechanical artifacts. Finally, reflecting on the ambiguity of the end of Capek's play he draws attention to the close connections between our notion of life and the valuative dimension – the specific values that we aim to realize in a human life seem to be tied up with the classification of something as a life-form. While the play can afford an aporetic ending on the question of what robots need in order to be able to participate in our life-form, we are in a situation, Schmid points out, where we cannot just wait and see how things play out.

The common theme running through all the papers is an attempt to understand conditions of sociality and normativity in interactions between humans and robots, and the conceptual problems that result in trying to fit robots in the conceptual and normative frameworks that have evolved for understanding human sociality and interaction. We hope that the book will aid readers to gain a better understanding of the descriptive issues involved, and thereby acquire new perspectives on the ethical questions raised by social robotics.

References

Coeckelbergh, M. (2012). *Growing moral relations: Critique of moral status ascription*. Houndmills/Basingstoke: Palgrave Macmillan.

euRobotics aisbl. (2013). Strategic research agenda for robotics in Europe 2014–2020. Retrieved from http://www.eurobotics-project.eu. Accessed 14 Sep 2016.

Gunkel, D. (2012). *The machine question*. Cambridge, MA: MIT Press.

Misselhorn, C. (Ed.). (2015). *Collective agency and cooperation in natural and artificial systems: Explanation, implementation and simulation* (Vol. 122). Cham: Springer.

Nørskov, M. (Ed.). (2015). *Social robots: Boundaries, potential, challenge*. Farnham, Surrey: Ashgate.

Seibt, J. (2017). Robophilosophy. In R. Braidotti & M. Hlavajova (Eds.), *Posthuman glossary*. London: Bloomsbury.

Seibt, J., Hakli, R., & Nørskov, M. (Eds.). (2014). *Sociable robots and the future of social relations: Proceedings of Robo-philosophy 2014* (Vol. 273). Amsterdam: IOS Press.

Seibt, J., Hakli, R., & Nørskov, M. (Eds.). (2018, forthcoming). *Robophilosophy: Philosophy of, for, and by social robotics*. Cambridge, MA: MIT Press.

Sharkey, A., & Sharkey, N. (2012). Granny and the robots: Ethical issues in robot care for the elderly. *Ethics and Information Technology, 14*(1), 27–40.

Sullins, J. P. (2005). Ethics and artificial life: From modeling to moral agents. *Ethics and Information Technology, 7*(3), 139–148.

Sullins, J. P. (2006). When is a robot a moral agent? *IRIE: International Review of Information Ethics, 6*, 23–30.

Wallach, W., & Allen, C. (2009). *Moral machines: Teaching robots right from wrong*. Oxford: Oxford University Press.

Chapter 2
Towards an Ontology of Simulated Social Interaction: Varieties of the "As If" for Robots and Humans

Johanna Seibt

Abstract The paper develops a general conceptual framework for the ontological classification of human-robot interaction. After arguing against fictionalist interpretations of human-robot interactions, I present five notions of simulation or partial realization, formally defined in terms of relationships between process systems (*approximating*, *displaying*, *mimicking*, *imitating*, and *replicating*). Since each of the n criterial processes for a type of two-agent interaction $\mathfrak{I}$ can be realized in at least six modes (full realization plus five modes of simulation), we receive a $(6^n \times n) \times (6^n \times n)$ matrix of symmetric and asymmetric modes of realizing $\mathfrak{I}$, called the "simulatory expansion" of interaction type $\mathfrak{I}$. Simulatory expansions of social interactions can be used to map out different kinds and degrees of sociality in human-human and human-robot interaction, relative to current notions of sociality in philosophy, anthropology, and linguistics. The classificatory framework developed (SISI) thus represents the field of possible simulated social interactions. SISI can be used to clarify which conceptual and empirical grounds we can draw on in order to evaluate capacities and affordances of robots for social interaction, and it provides the conceptual means to build up a taxonomy of human-robot interaction.

Keywords Social ontology • Classification • Social robots • Sociality • Simulated interaction • Taxonomy • Human-robot interaction

---

J. Seibt (✉)
Research Unit for Robophilosophy, Department for Philosophy and History of Ideas, School of Culture and Society, Aarhus University, Denmark Jens Chr. Skous Vej 7, DK-8000, Aarhus C, Denmark
e-mail: filseibt@cas.au.dk

R. Hakli, J. Seibt (eds.), *Sociality and Normativity for Robots*, Studies in the Philosophy of Sociality 9, DOI 10.1007/978-3-319-53133-5_2

2.1 Introduction

At first blush, the notion of a "social robot" does not seem to make much sense. If we use prototype semantics to interpret the terms involved, we end up with a category mistake—the currently prototypical robot, i.e., the industrial robot performing repetitive movements irrespective of context, is not the kind of entity that could engage in prototypical social interactions such as reaching an agreement or meeting for a cup of coffee. If we set prototype semantics aside and turn to explicit definitions, matters do not seem to improve at all. For, on the one hand, researchers in "social robotics" advise us to abandon all hope with regard to conceptual fixations within a fast changing field—"the concept of robot is a moving target, we constantly reinvent what we consider to be 'robot'" (Dautenhahn, 2014). On the other hand, while the term "social" still is taken to be amenable of definition, so far there is only a research path *towards* a comprehensive account of human sociality but not yet a theory with stabilized terminology. Research on social ontology in philosophy is yet to come into full contact with efforts by anthropologists, linguists, psychologists, and cognitive scientists to create an integrated study of human sociality (Enfield & Levinson, 2006). In short, we have only begun to determine, based on conceptual and empirical research, what could or should be understood by the qualifier "social," and the class of items we wish to qualify with this yet rather vague predicate—the reference class of "robots"—is under construction, even constitutively perhaps.

This is an unusual situation. The philosopher's uneasiness can only increase, however, upon reading that a "social robot" is designed to enable people to "interact with it as if it were a person, and ultimately as a friend" (Breazeal, 2002, p. xi). This one-liner nicely encapsulates three assumptions that, from the philosophical point of view, raise conceptual issues of breath-taking significance.

The first issue concerns the idea that human-robot interactions could qualify as instances of *social* interactions—are we justified in applying the term "social interaction" even when the relevant capacities for sociality are not distributed symmetrically over the interacting systems? In short, could there be, in principle, *asymmetric* social interactions?

The second issue centers on the type of capacities that robotic systems could possibly exhibit. Given the current state of robotics, most of those capacities that we consider distinctively human, such as linguistic communication and intelligent behavior, are merely approximated by robotic systems. But can capacities for sociality at all be *approximated*? Which degrees and kinds of approximation are admissible, and can robots have the relevant approximative capacities?

The third issue concerns differences in the normative dimension of classificatory predicates. The predicates (i) *x is a y* and (ii) *x can be interacted with as if it were some y* carry different semantic restrictions and signal different practical implications. Will the practice of human-robot interaction change the semantics and normative significance of ascriptive predicates (e.g., *x is a person*), i.e., will we come to allow for anything normatively to count as a person if our interactions with it resemble interactions with a person?

The philosophical debate about social robotics so far has concentrated on ethical and moral aspects, either sidestepping conceptual issues or addressing them obliquely within a specific context of application. This focus made good sense, given that the basic design goal of social robotics – artifacts that engage humans in interaction types that we used to reserve for humans only – generates a host of urgent questions of policy and legislation. Up until recently philosophy of social robotics mainly engaged in "roboethics" (Veruggio, 2006) understood as an area of applied ethics that discusses benefits and disvalues of social robotics applications in elderly care, schools, or therapeutic contexts (cf. e.g. Sullins, 2008; Sharkey & Sharkey, 2012; Sparrow & Sparrow, 2006; Coeckelbergh, 2010; Vallor, 2011).

Increasingly, however, the scope of roboethics has widened to take up general methodological issues, exploring the challenges of social robotics to traditional paradigms of moral agency and moral patiency (cf. e.g. Gunkel, 2012; Coeckelbergh, 2012). As these investigations bring out clearly, supported by empirical research on human-robot interaction, the phenomenon of human-robot interaction contains conceptual ambiguities that are directly linked to ethical judgments (cf. e.g. Kahn, Freier, Friedman, Severson, & Feldman, 2004).

We need to conceptually clarify the phenomenon of human-robot interactions in all its diversity, in order to make progress in the professional and public ethical debate about social robots. But which methods should we apply if the traditional essential differences between *res extensa* versus *res cogitans* are dissolving? Postphenomenological philosophy of technology recommends the radical step of abandoning ontology. However, there are good epistemological reasons, in my view, why roboethics should rely on both empirical research and ontology to provide a fine-grained analysis of human-robot interactions. Since public policy and legislation are guided by ethical recommendations, it is problematic if ethical recommendations are based on subjective impressions about what a robot "is" or "is not"—it may be presumptuous, for example, to suggest without more detailed ontological and empirical investigations that human interactions with carebots are "too undignified for a human to enter into" since carebots are "totally inadequate surrogate products" (Danish Ethical Council, 2010). There is increasing empirical evidence that in certain contexts of therapy and caretaking (e.g., autism and dementia) humans benefit more from the interaction with robots than from human interaction (Cabibihan, Javed, Ang Jr, & Aljunied, 2013; Leyzberg, Avrunin, Liu, & Scassellati, 2011).

More generally speaking, the Western democratic conception of political authority is grounded in the epistemology of discourse-theoretic validation; ethical judgements are to be warranted by "rational discourse," which in turn presupposes normative classifications that are established by intersubjective methods of empirical and theoretical inquiry. When postphenomenologists dismiss "ontology" in favor of a postphenomenological or "relational" approach focused on subjective "takings-as," they miss out on an important ally that can translate the "relational approach" into the conceptual format needed for the sort of rational discourse that in Western societies still counts as epistemic warrant for ethical and political decision making. While—correctly, in my view—ridding themselves of the murky bathwater of traditional realist "substance–attribute" ontology, postphenomenologists inadvertently throw out the baby of pragmatist analytical ontology, which alone is currently capable of developing sufficiently differentiated and intersubjectively justifiable normative classifications of interactions.[1]

The following considerations, written from the perspective of pragmatist analytical ontology, are intended to offer clarifications and classificatory tools that should prove useful for the ethical debate about social robotics applications. In addition, the conceptual distinctions offered below may also have heuristic value for the design of empirical studies of human-robot interactions, especially when undertaken in the context of research on social cognition. I will outline a conceptual framework on the basis of which any extant and future type of human-robot (and human-human) interaction can be classified, so that a comprehensive and detailed taxonomy of human-robot interactions can be built up incrementally.

I begin in Sect. 2.2 with a brief reflection on the predication contexts "*x* treats *y* as *z*," "*x* interacts with *y* as if it were *z*," and "*x* takes *y* as *z*." As I explain, these contexts not only govern different modes of constructionality (make-believe, fictionality, conventionality or "social reality") they also behave differently with respect to precisely those ascriptive terms that social roboticists are wont to use in their descriptions of human-robot interactions. I argue that for conceptual reasons we cannot adopt the—temptingly easy—strategy of treating human-robot interactions as fictionalist analogues to human-human interactions. Thus we are saddled with what I call the "soft problem in the ontology of social robotics," namely, the problem of how to describe human-robot interactions, from a second-person and third-person point of view, given that our concepts for human social interactions as such are inapplicable. I suggest addressing the soft problem by developing a theory of simulated social interaction. In Sect. 2.3 I set out on this task and define five notions of simulation, formulated as relationships between process systems: replicating, imitating, mimicking, displaying, and approximating. In Sect. 2.4 I sketch how these distinctions can be used to define for each interaction concept $\mathfrak{I}$ a "simulatory expansion" of $\mathfrak{I}$ and show how the latter can be used to characterize human-robot interactions. In Sect. 2.5 I address the question of which sorts of simulations of a social interaction $\mathfrak{I}$ can qualify as a social interaction. Given that we currently have not one but several competing notions of sociality, I suggest that we should abandon the idea of a dualist distinction between social and non-social interactions; rather, we should conceive of sociality as a matter of degree. I show how we can use the notion of a simulatory expansion of a social interaction $\mathfrak{I}$ to create a taxonomy for human-robot interactions. As I shall explain, this taxonomy promises to enable us to address two tasks in one go. On the one hand, it should be possible to integrate within this taxonomy competing accounts of sociality with more and less restrictive requirements. On the other hand, with the descriptive tools of the taxonomy we

[1]Note that pragmatist analytical ontology is not committed to the facticity or even the possibility of rational discourse, just to its utility as regulative idea and regulated praxis.

can create a comprehensive conceptualization of simulated sociality that allows for differentiated descriptions of human-robot interactions from a second-person and third-person point of view.

2.2 Against Fictionalist Deflations of Robot Sociality

Social robotics creates a practical space that we currently have great difficulty conceptualizing. While adults interacting with social robots display signs of conceptual disorientation, children tend to resolve the conflict pragmatically by using contextual categorizations of robots ("alive enough to be *x*"), or hybrid categorizations ("mechanical yet with feelings," cf. Bernstein & Crowley, 2008), or practically relate to robots in ways that conflict with the chosen categorization (cf. Clark, 2008; Turkle, 2011). While most researchers might agree that we need a new ontological category (cf. Kahn, Friedman, Perez-Granados, & Freier, 2004) in order to conceptualize human interactions with "social robots" so-called, it is currently far from clear how we should go about this task. Do we need a new concept for the item we are interacting with, the robot, as a new sort of artificial "living thing"? Or, leaving the robot in the familiar category of machines, do we need a new concept for our responsive dispositions for interacting with such machines, e.g., as dispositions for machine sociality? Or should both relata, robot and human, retain their traditional conceptualizations and should we introduce a new sort of relation between them: fictional social relations?

These are three basic strategies that traditional substance–attribute ontology has to offer, as well as various combinations thereof. Once we abandon the traditional substance paradigm, however, new systematic options for categories come into view. In this paper I will suggest quite a different line of approach, however. It is not *one or two* new ontological categories that we need in order to conceptualize human-robot interactions, I shall argue, but an entire new classificatory framework for simulated social interactions. In order to motivate this approach I will begin with some general remarks on the idea that human-robot interactions could be *fictional* social interactions.

Humans interact with their environment not only (i) physically but also (ii) symbolically, i.e., representing physical features, and (iii) "figuratively," i.e., assigning to representations of physical features new interactive significances. Consider the following three types of "figurative" interactions of a human with her or his environment (an object, another human, an event etc.),[2] which I introduce here together with their characteristic linguistic forms, i.e., linguistic expressions that are typically used in descriptions of the participant, who experiences the interaction

[2]For the sake of simplification I shall throughout this paper assume that an interaction has just two participants (i.e., two human participants, or a human and a robot).

from a second-person point of view, and descriptions of an observer from a third person point of view, respectively.

1. *Make-believe or pretend-play*. The participant describes the object she is interacting with according to the interpretational rules of the make-believe scenario (e.g., "I tied my horse to a branch"), and the observer describes the interaction as make-believe scenario extrapolating from normal interpretational rules based on similiarity (e.g., "she treated the stick *as* a horse").
2. *Fictional interaction*. The participant describes the interaction in accordance with the conventions of the fiction (e.g., "I greeted the king and was admitted to the court"), the observer describes the participant's behavior together with its significance relative to the conventions of the fiction (e.g., "she bowed as if she were greeting a king and he moved his head and right hand as if he were admitting her to approach").
3. *Socially instituted interaction*. The participant describes the interaction in accordance with extant social conventions (e.g., "I showed him the receipt") and the observer describes the participant's behavior and its social significance (e.g., "she handed him a piece of paper that counts as proof of payment").

Observer descriptions of make-believe scenarios are linguistically typically signaled by the phrase "x treats (considers) y as z (or: as if it were z)," while fictional interactions are typically described by "x interacts with y as if it were doing z," and socially instituted interactions are typically characterized by the phrase "x takes (object or interaction) y to count as z."[3]

With these distinctions in mind, let us now consider the following formulations of the design goals of social robotics and descriptions of human-robot interactions, respectively:

1. "We interact with [a sociable robot] as if it were a person, and ultimately as a friend" (Breazeal, 2002, p. ix).
2. "Ideally, people will treat Kismet as if it were a socially aware creature with thoughts, intents, desires, and feelings. Believability is the goal. Realism is not necessary" (Breazeal, 2002, p. 52).
3. "This also promotes natural interactions with the robot, making it easier for them to engage the robot as if it were a very young child or adored pet" (Breazeal, 2002, p. 100).
4. "I find people willing to seriously consider robots not only as pets but as potential friends, confidants, and even romantic partners" (Turkle, 2011, p. 26).
5. "…social robots—the class of robots that people anthropomorphize in order to interact with them" (Breazeal, 2003, p. 167).

Formulations (2) and (4) describe human-robot interactions as make-believe scenarios where an *object* is treated as something else, or as if it were something

[3]For the sake of the argument in this section I operate here with a simplified version of Searle's definition of social reality: "For all kinds Z, instances of kind Z is part of social reality iff there are X, Y, and C: X takes Y to count as a Z in circumstances C," (cf. Searle, 2010).

else. Humans mainly engage in make-believe projections during childhood (pretend play), but many people allow themselves to entertain the special type of *anthropomorphistic* make-believe projections also in adult life, as a form of conscious, self-ironic sentimentality. Formulation (5) explicitly clarifies that the design of social robotics consciously targets our capacities and dispositions for anthropomorphizing make-believe projections.

Very little indeed is needed to anchor such projections and to allow us to treat an object as a human being or as a companion. You may treat the tree in front of your window as the master of the garden just because it overshadows all other plants of the garden, and you may treat your car as your companion or adversary just because its start-up performance in difficult weather resembles reliable or malicious actions.[4]

Projections of make-believe ("treating *x* as if it were *y*") are based on physical or functional similarities or analogies between, on the one hand, features or doings of an intentional agent, and, on the other hand, static or dynamic features of a natural or artificial item (Walton, 1990). Fictional interactions are also based on similarities or analogies, but there is an important difference between "treating some *x* as if it were to do *y*" and "interacting with *x* as if it were *y*." When I treat my car as companion and greet it, the car does not perform any distinctive behavior in return; in contrast, a NAO-robot reacts to a human greeting with greeting behavior, or even autonomously displays greeting behavior to elicit a human greeting. In other words, make-believe scenarios typically are one-sided analogical projections where only the human agent involved executes the actions in the template of actions and reactions that define the interaction. In contrast, in fictional interactions both agents[5] behave in ways that resemble the actions and reactions prescribed by the interaction template. Typically we connect fictional interactions with role-play, where the relevant fictional conventions are understood by all agents involved. But a fictional interaction can also be said to take place even if one of the agents is not conscious or is not aware of any convention of fictionality being in place. For example, as long as the behavior of my dog resembles a greeting or as response to my greeting, one can say that I interacted with him as if we were greeting each other; and as long as the behavior of a hypochondriac resembles pain relief after medical treatment, one can say that by administering a placebo I interacted with him as if he were in pain. In short, whether a scenario is make-believe or a fictional interaction

[4]Here I bracket the question whether "anthropomorphizing" is the right label for make-believe projections of this kind. Treating something as companion or foe does not necessarily imply treating it as human being. Especially if one applies the "non-exceptionalist" notions of sociality I discuss below, one might argue that even though human beings are the primary instances of social actors, our long-standing practice of projecting social roles onto natural things and artifacts is a way to *"socialize"* the world, not to "anthropomorphize" it.

[5]Throughout this paper I use the term "agent" in the broad disjunctive sense where it refers either to agents proper, i.e., conscious living things that can act on intentions, *or* to inanimate or living items that are causal origins of events that resemble (and thus can be treated as) actions, i.e., the doings of agents proper.

depends on whether the agents engender occurrences that resemble the actions and reactions of an interaction template.

Applying these considerations to quotations (1) through (5) above, human-robot interactions are described as make-believe in (2), (4) and (5) (here explicitly), and as fictional interaction in (1) and (3) (in the context of this passage Breazeal explains that her robot Kismet is programmed to produce behavioral patterns that make it more "natural" for people to interact with it "as if it were a very young child or adored pet"). None of the quoted announcements of the design goals of social robotics use formulations that are the characteristic indicators of social actions, i.e., the movements of the robot are not supposed to be "taken as" or to "count as" certain actions, nor are the participating humans said to exhibit behavior that "counts as" an action with social significance. In other words, the quoted passages do not describe human interactions with robots as scenarios where a social action *de facto* occurs.

In comparison with other descriptions of the design goals of social robotics in terms of intentionalist vocabulary, where robots are said to "perceive" their environment or even are said to be "aware" of it (cf. Fong, Nourbakhsh, & Dautenhahn, 2003, p. 145), the strategy of using linguistic forms that express the fictional irrealis of sociality rather than the realis may seem a very useful device of professional caution. But upon a closer look any attempt to deflate the question of robot sociality by using social vocabulary in fictionalist embeddings are bound to be unsuccessful, as I shall argue now.

Consider again formulation (1). Could we ever interact with anything *"as if it were a person"*? As just mentioned, both make-believe and fictional interaction are based on resemblances or analogies between descriptive aspects of entities and interactions. But the predicate "person" is not a descriptive predicate. When we call an entity a person, we thereby make certain commitments in the performance of that very utterance—we are not describing features but announce that certain commitments are undertaken. The performative-ascriptive use of language is not limited to promises and explicit declarations—it pervades our vocabulary for social interactions. Importantly, performative-ascriptive predicates cannot be embedded in contexts with fictionality markers. One cannot perform a linguistic-pragmatically and conceptually coherent speech act by uttering "It is as if I hereby promise you ...," nor "what I will say now is a bit like a promise ... " Similarly, if we treat some x as a person we are committed to taking x to count as a person—that is, assuming that we wish to abide by linguistic norms and the actions they entail, we *must* interact with x *as* a person.

This is due to two facts of social reality.[6] The first fact is that commitments are strictly "bivalent"—they are either undertaken or they are not undertaken; pretending to undertake a commitment is simply *to fail* to undertake it. But if we cannot make fictional commitments, we cannot make fictional promises, nor can we

[6]In the context of this paper I take it that these two facts are self-evident elements of the "logic" of social practices; a more detailed discussion of the semantics of fictional discourse in application to the performative-ascriptive predicates for social and moral roles is in preparation.

interact with someone as if she were a person. At best we can commit ourselves to treat an occurrence as having a certain special significance within a fictional context, but this is a real commitment to abide by special interpretational conventions.[7] Making a promise or treating someone as a person are real social interactions *by virtue* of engaging in a certain declarative practice (e.g., by uttering a sentence) which *is* the process of taking on the relevant commitments. If we address an infant as a person, for example, we are making a full commitment to certain actions and omissions, in accordance with rights and obligations.[8]

The second, more general reason for why one cannot make fictional commitments are the performance conditions of social interactions. It is a constitutive feature of social interactions that the agents involved can only react to the enactment of behavior since the other agent's intentions as such are not directly accessible. The agents of a social interaction $\mathfrak{I}$ take certain behaviors (their own and that of others) *to count as* social actions of a certain kind (e.g., a greeting, answering, helping) and thus to *imply* that the agents involved have the specific intentions associated with $\mathfrak{I}$. But the factual occurrence of these intentions is not part of performance conditions for $\mathfrak{I}$. To be sure, the action of greeting and the action of pretending to greet are two different actions—but in the context of a social interaction the behaviors involved in each of these actions *count as* the same *social* action, since they both fulfill the relevant performance conditions for the *social* action of greeting. To emphasize, I am not prejudging here the issue of how to define sociality, i.e., which sorts of joint or reciprocal acts of intentionality should occur for a behavior to count as social interaction $\mathfrak{I}$ in context *C*. I am making the pedestrian point that it is an obvious requirement of social praxis that the *performance conditions* of a social interaction—as opposed to its definition—must relate only to behavioral criteria and cannot take intentional states into account.[9] Often we criticize that a social

[7]A promise given by a fictional character of a stage play—e.g., Romeo's promise to Juliet to return the following night—is a behavior that counts as promise with respect to our actual social conventions. But due to the referential shifts introduced by the context of the stage play, the actor playing Romeo makes no commitment at all beyond the commitment to playing Romeo—the commitment to return is not the actor's fictional commitment but the commitment of the fictional character Romeo. It is an act that in the context of the fiction counts as the promise of the fictional character *x* that *p*—and in this the only coherent sense, I think, we can give to the idea of a "fictional commitment."—My views here benefitted from many discussions with Stefan Larsen, in the course of his PhD project on the status of robot sociality, see Larsen (How to Build a Robot and Make It Your Friend. PhD dissertation. Aarhus University, Denmark, 2016); Larsen offers a detailed investigation of the fictionalist conventions of the theatre and uses them as heuristic metaphor for the description of human-robot interactions.

[8]By contrast with the infelicitous formulation in (1), when Breazeal describes a colleague's interaction with his infant daughter she keeps the fictional scope correctly focused on the *descriptive* predicate "socially fully aware and responsive [agent]": "Simply stated, he treats his infant as if she is already fully socially aware and responsive—with thoughts, wishes, intents, desires, and feelings that she is trying to communicate to him as any other person would" (Breazeal, 2002, p. 30).

[9]Unless one champions a purely behaviorist account of social interactions, the definition of a social interaction will include behavioral performance conditions, but also state additional conditions relating to the agent's intentions and understanding of norms.

interaction was performed without the intentional state that according to the concept of the interaction should have been part of it—we complain that someone merely "went through the motions" in his greeting or offered an insincere apology, etc. Such charges about someone's performing a social action inauthentically or insincerely relate to the *quality* of the social action, not to the latter's *occurrence*—we do not question that the social interaction has been performed, but claim that it has been performed badly.[10]

In combination the mentioned two facts of social reality—i.e., that commitments are binary and that the performance conditions for social interactions relate to norm-governed behavior and not to agentive intentions—preclude the possibility of fictional social interactions. A fortiori, they rule out that declarative-performative terms can be used in fictional contexts.

In other words, for social interactions *there is no fictionality gap*. Fictionality requires that the real and the fictional domain are in some fashion separated—that there is a "gap" between the two domains that is bridged by imaginative analogical reasoning, using the real as stand-in for the fictional (Walton, 1990, 2005). We may treat a stick as a horse or interact with a robot as if it were a "young child" and yet remain fully aware of the difference in descriptive features between the real stick and the imagined horse, or the robot and the imagined child, respectively. But no such fictionality gap can be kept open for social actions and interactions. This is most palpable when performative-ascriptive predicates are used or implied—whoever addresses some x as a person, by that very utterance undertakes the relevant commitments and thus takes x to count as a person, i.e., engages in a real social interaction. It is important to see, however, that the fictionality gap collapses not only for performative-ascriptive predicates but for all predicates for social actions and thus obviates a fictionalist interpretation of human-robot sociality in general.[11] From a philosophical viewpoint it is thus a category mistake to assume that we can interact with anything "as if it were a person." The fictionalist interpretation of human-robot interactions overlooks that we cannot drag social (inter-)actions into the scope of fictionality. We can create contexts of social interaction for fictional interactions, i.e., we can commit ourselves to respect the reinterpretations of a fictional domain,

[10]Since the performance conditions for social interactions relate only to behavior and not to the intentions of the agents involved, behaving as if one were to perform a certain social action A that is part of a social interaction $\mathfrak{I}$ is tantamount to exhibiting behavior that counts as the relevant part of social action $\mathfrak{I}$. The agent might intend to perform another action, e.g., she might intend to do B = [pretend doing A], but this does not detract from the fact that the performance conditions for A have been fulfilled.

[11]In Seibt (2014b) I use the contrast "friend" vs. "person" to highlight the difference between descriptive and performative-ascriptive terms—in hindsight, this was infelicitous, since even though the term "friend" is partly descriptive, the performative-declarative elements of the meaning of friendship arguably are dominant. The predicates "friend" and "person" belong on the same side and should be contrasted with (predominantly) descriptive predicates such as "child" or "woman."

but we cannot fictionalize our commitments. In short, there is social fictionality but no fictional sociality.[12]

To summarize the observations in this section, I have argued here that we cannot treat human-robot interactions as fictional social interactions, i.e., as interactions where humans engage in fictional commitments to a social interaction. While for physical interactions there is a clear difference between the real and the fictional interaction (e.g., pushing someone and behaving as if one were pushing someone) due to physical dissimilarities, for social interactions the fictionality gap collapses. A human's actions towards the robot cannot be considered as a pretended social action, since the performance conditions of social actions, i.e., the conditions for the performance of a social action from a third-person point of view, are insensitive to the difference between authentic and pretended action. As convenient it might seem to describe human interactions with a robot as fictional social actions, upon a closer look such descriptions are conceptually incoherent.

This leaves us with what I call the "soft problem in robo-ontology." Once we realize that we cannot use fictionalizations of familiar predicates for human social interactions in order to describe human-robot interactions, we face the following juncture. Extant philosophical definitions of social interactions determine that for a social interaction to occur, all agents involved must have the capacities required for normative agency, e.g., intentionality, consciousness, normative understanding.[13] Thus, going by our standard philosophical definitions, we need to treat human interactions with so-called "social" robots as non-social interactions. Alternatively, we need to develop new conceptual tools for forms of non-reciprocal or asymmetric sociality, i.e., for social interactions where one agent lacks the capacities required for normative agency.

In the following sections I will pursue the second route and outline a solution to the "soft problem" by devising a conceptual framework for simulated social interactions. By analyzing forms of simulations and classifying types of interactions parts of which are simulated, we can investigate in a differentiated fashion which, if any, human-robot interactions can qualify as social interactions. As will become

---

[12]That we cannot uphold a fictionality gap for social interactions is also implicitly reflected in concerns about the expectable cultural change effected by social robotics. Authors who warn against the pervasive use of social robots are not worried about humans losing themselves in realms of fictionality. Rather, they fear a *degradation* of social interactions, due to an increased functionalization of social relations (cf. Sharkey & Sharkey, 2012) where "the performance of connection seems connection enough" (cf. Turkle, 2011, p. 26). The worry is that we increasingly will reduce to definition of a social action to its performance conditions, i.e., that we will abandon our current understanding that the concept of a social action comprises both conditions relating to behavior and conditions relating to intentional states. While we currently criticize each other for performing bad social actions by going through the motions, by not living up to the concept of the social action in question, such criticism will vanish—or so the argument goes—and with it the social standards for performing social actions well, i.e., sincerely.

[13]For an overview of these requirements of standard accounts of social interaction see Hakli (2014). Hakli discusses what I call "the soft problem" in a differentiated fashion that also involves larger epistemological perspectives on conceptual change and the conventionality of conceptual contents.

apparent, I hope, discussing robot sociality in the terms of the suggested framework can also advance our understanding of sociality in general.

2.3 Five Notions of Simulation

The "soft problem of robo-ontology" introduced in the previous section is best addressed, in my view, by devising a conceptual framework that is rich and discriminative enough to describe human-robot interactions at all stages of research development in robotics. The first step towards developing such a framework is to clarify the notion of simulation.

As we common-sensically understand the term, the predicate "simulate" states a relationship between occurrences. In order to describe the denotation of the predicate more precisely, I suggest using a process ontology, i.e., an ontology that countenances processes among the basic categories. The following definitions are drafted on the background of a mono-categoreal process ontology called *General Process Theory*, but in the context of this paper it can remain open whether the embedding ontological framework has other basic categories, and how the explanatory predicates of the ontological theory (e.g., "emerges," "is part of," or "is an instance of") can be defined more precisely.[14] Since in General Process Theory processes are non-particulars, I shall use "process" and "process type" interchangeably.

Let "C" and "C^*" stand for processes or collections of processes. A collection of processes can occur in such a way that (i) other processes "emerge" (in the period during which the collection occurs) or else (ii) without such "emergent products." Thus we can distinguish two sorts of co-occurrent collection of processes, (i) *complex processes* and (ii) *collections of processes*. Both complex processes and collections of processes may have themselves complex processes (or collections of processes) as parts. If a process system Σ is a collection of processes, the occurrence of Σ's parts or subsystems C and C^* is merely a mereological sum, i.e., $\Sigma = \text{sum}(C, C^*)$. In contrast, if a process system Σ is a complex process, then there is a basic process, the interaction I of C and C^* that emerges from the interference of C and C^*, i.e., $\Sigma = I(C, C^*)$.[15] Both complex processes and collections of processes may be parts of complex processes or of collections of processes.

The distinction between a complex of processes and a collection of processes will become relevant below when we consider alternative classifications of processes. For example, if Kim works on a production line next to an industrial robot, she might take herself to work *alongside* of the robot, treating the robot as a feature of the production line, or she might take herself to work *with* the robot in a collaborative

[14] Relevant technical details, especially also on the part-relation for processes I shall use in the following definitions, can be found in Seibt (2005, 2009, 2014a).

[15] For details cf. Seibt (2014a).

situation. This difference will matter when we discuss differences in classifications of social interactions from first, second, and third-person points of view.

In order to allow for such further distinctions the following definitions for modes of simulations are formulated in such a way that they can be applied to process collections with arbitrary dynamic organization. That is, the following relationships may hold for process systems Σ and Σ^* if these are parts of a mere collection of processes, occurring "alongside," but also if Σ and Σ^* are parts of a complex process, i.e., parts of an interaction.

The five relationships of simulation I will distinguish here are five ways of deviating from a common baseline, which I call the "realization" of an action[16]:

Definition 1 (Realization) For any process type A, process system Σ realizes A iff Σ generates an instance of A.

For present purposes, let us say that a process system Σ *generates* a process α just in case α is a (possibly emergent) part of Σ, and that α *is an instance of* process type A just in case α fulfills all (functional and non-functional) occurrence conditions F_i for A.[17] For example, two people reading next to each other each realize instances of the action type *reading*.

Any two instances of a process type A differ with respect to at least one aspect that is not among the occurrence conditions F_i for A, e.g., its location. If a process α fulfills all functionally relevant occurrence conditions of A but fails to fulfill at least one of the non-functional occurrence conditions of A (e.g., conditions of A's biochemical realization), α is no longer an instance of A but an instance of a process type that is functionally equivalent to A. Deviations from non-functional occurrence conditions are the hallmark of the first and strongest mode of simulation.

Definition 2 (*SIM-1*, Functional replication) For any process type A_P realizable by process system Σ, process system Σ^* functionally replicates A_P iff Σ^* realizes a process type C that is distributively functionally equivalent to A_P, or Σ^* realizes

[16]The following five definitions are simplified—and, I hope, thereby improved—versions of those presented in Seibt (2014b). In the following I use capital letters as variables for process types and Greek letters as variables for instances of a process type. Note, though, that in *General Process Theory* there are, strictly speaking, no "instances" or "tokens", since I consider the type-token (kind-instance) distinction to mark extremal regions on a gradient scale of specificity; in order to simplify the exposition here I stay with the traditional idiom and speak of highly specific, localized processes as instances (tokens) of kinds (types) of processes.

[17]For present purposes I must rely on an intuitive understanding of the theoretical predicates "functional" and "non-functional"; a more precise statement of the envisaged distinction is quite involved, especially from within a naturalist metaphysics where all properties, even *qualia*, are in some sense "functionalized."— Here I am also neglecting specific issues of realizations that arise when the process in question is an action and intentional scope comes into play. In Seibt (2014b) I formulate the right-hand side of the biconditional as a disjunction: action A is realized by a process system Σ if the system generates an instance of A *or* if Σ realizes instances of all subprocesses $B_1, \ldots, B_n$ of A. The disjunctive formulation is to account for variations in intentional scope, e.g., the difference between the holistic performance of an action by an expert versus the summative performance of the same action by a novice.

subprocesses $D_1, \ldots, D_n$ of C that are distributively functionally equivalent to all subprocesses $B_1, \ldots, B_n$ of A_P.[18]

Let me unpack this definition. First, the disjunctive clause in the righthand side of *SIM-1* is to ensure that the definition covers simulations of and by (i) simple processes, (ii) complex processes, as well as (iii) collections of processes. Second, processes X and Y are *distributively functionally equivalent* iff for each part of X there is an input-output equivalent process that is part of Y. Third, the parthood relation on processes is not the transitive parthood relation that holds for spatiotemporal regions, i.e., it is not the case that every part of the spatiotemporal region occupied by a process X is a part of X. Rather, parthood on processes is a non-transitive and context-dependent relation (Seibt, 2014a). Claims about parts of a process A are thus always to be understood in relation to a finite partition P that states the mereological structure of A. In order to highlight that claims about functional replication—and other modes of simulation—are relative to a given finite partition of the simulated process type, a subscript "P" is added to the variable of the process type, i.e, "A_P". For example, relative to a fine-grained partition for the process type A = *teaching math*, only an imaginary close-to-perfect android such as *Star Trek's* Mr. Data could functionally replicate A = *teaching math*; on the other hand, relative to a coarse-grained partition of A = *initiating a conversation* one can claim that even present-day educational robots such as Aldebaran's NAO-robot functionally replicate A.

The definition of functional replication demands maximal similarity of functional structure (to reemphasize: relative to a given partition of the simulated process type). The second mode of simulation, here called "imitation," loosens this constraint by reducing the requirement of distributed input-output equivalence, i.e., input-output equivalence along all parts of the simulated process, to a requirement of input-output equivalence for most parts. Let us say that processes X and Y are *functional analogues* iff all non-ultimate parts of X have functional equivalents in Y (where a "non-ultimate" part of X is any part that has a part in the partition of X). In other words, processes X and Y are functional analogues just in case they are input-output equivalent and have similar (though not maximally similar) functional structure. Aiming again, as in *SIM-1*, for a definition that can cover simple and complex processes as well as collections (complexes) of processes, we can define a second mode of simulation as follows:

Definition 3 (*SIM-2*, Imitation) For any process type A_P that has non-ultimate parts and is realizable by process system Σ, process system Σ^* *imitates* A_P iff Σ^* realizes a process type C that is a functional analogue of A_P, or Σ^* realizes subprocesses $D_1, \ldots, D_n$ of C that are functional analogues of all immediate subprocesses $B_1, \ldots, B_n$ of A_P.

[18]More precisely the definiendum should be formulated as: "process system Σ^* functionally replicates A_P as realizable in Σ"; here and in the following definitions I take this restriction to be understood.

For example, let us assume that the partition for the process A = *fetching new linen when commanded* lists among the parts of A the process B_1 = *reacting to a command* which has as non-ultimate parts B_2 = *classifying type of speech act* and B_3 = *analyzing semantic content*. Relative to this partition, Aethon's TUG robot, a mobile, self-navigating delivery device for hospitals that is activated by touchscreen input, cannot be said to functionally replicate process *fetching new linen when commanded*, since TUG does not generate processes that are functionally equivalent to B_1, B_2, and B_3. But TUG generate a process that is functionally analogous to B_1, B_2, and B_3, and thus can be said to imitate A. Or again, to illustrate the second disjunct in *SIM-2*, a multi-robot system of Kilobots imitates the group process A = *forming a V-shape* since the immediate subprocesses of A as realized by, e.g., a swarm of birds, i.e., the local interactions between neighboring birds, have functional analogues in the programmed interactions between the kilobots.

The third mode of simulation, which I call "mimicking," further relaxes the required input-output equivalence of simulating and simulated process. For the purposes of social robotics, one might argue, the first two modes of simulation are not really relevant since it does not matter whether two processes are input-output equivalent in all regards—all that matters is input-output equivalence relative to the capacities of human observation. Let us say that process Y is an *empirical proxy* of X iff for any observable part of X there is a part of Y that is observably input-output equivalent. A process may then be said to be an empirical proxy of another even if there are considerable deviations in functional structure.

Definition 4 (*SIM-3*, Mimicking) For any process type A_P that has observable parts and is realizable by process system Σ, process system Σ^* *mimicks* A_P iff Σ^* realizes a process type C that is an empirical proxy of A_P, or Σ^* realizes subprocesses $D_1, \ldots, D_n$ of C that are empirical proxies of all immediate parts $B_1, \ldots, B_n$ of A_P.

For example, Hiroshi Ishiguro's *Geminoid* robots, which are physical copies of the bodies of particular people, can be said to mimick some human facial movements, e.g., rolling one's eyes or smiling, since—or so we may assume for the sake of the illustration—these artificial facial movements are empirically indistinguishable, down to their smallest observable phases, from their real counterparts in a human face.

By ways of contrast, consider Cynthia Breazeal's robotic head *Kismet*. Kismet also moves some of its components (called Kismet's "ears," "lips," "eyebrows," and "eyelids") in ways designed to be interpretable as expressions of surprise, delight, disgust, sadness, etc., but here the observable similarity with human expressions of these emotions is reduced to some typical elements. Kismet performs the facial movements familiar from cartoons and animations, and like these it exploits our tendency to "anthropomorphize" (or better: *socialize*) our environment.[19] When

[19]As noted above, Footnote 4, instead of "anthropomorphizing" we should rather speak of a human tendency to "socialize" the environment. Above I pointed out that the performance conditions for

Kismet and similar social robotics applications are designed in ways that capitalize on the human tendency to socialize their environment, the following mode of simulation is targeted.

Definition 5 (*SIM-4*, Displaying) For any process type A_P that has parts and is realizable by process system Σ, process system Σ^* *displays* A_P iff Σ^* realizes subprocesses $D_1, \ldots, D_n$ of C that are empirical proxies of all and only typical immediate parts $B_1, \ldots, B_n$ of A_P.

Which parts $B_1, \ldots, B_n$ of a process A are "typical" for A in the sense that a human being will take the occurrence of empirical proxies of these parts as an indication for the occurrence of A I take to be context-dependent; but note that some relevant aspects of Gestalt perception may be innate and some typical cues for behavior that steer the dynamics of social interactions, such as pointing and gaze-following, may be culturally universal (cf. Enfield & Levinson, 2006). Thus claims about the *displaying* of a process A in the sense of *SIM-4* are relative to a partition P whose nodes are annotated for typicality.

Displaying is the target mode of simulation for social robotics applications that belong into the genre of entertainment technology, where smooth social interactions are instrumental for the primary design goal of engaging the user in a game. For example, Sony's robotic dog *AIBO* can be said to display a dog's greeting or a dog's invitation to play.

Other applications, especially in the area of assistive technology where special cooperative efforts on the side of the human can be expected, may settle for the fifth and "poorest" mode of simulation:

Definition 6 (*SIM-5*, Approximating) For any process type A_P that has parts and is realizable by process system Σ, process system Σ^* *approximates* A_P iff Σ^* realizes subprocesses $D_1, \ldots, D_n$ of C that are empirical proxies of only some typical immediate parts $B_1, \ldots, B_n$ of A_P.

For example, the robotic seal PARO approximates process type A = *a cat's purring* by creating empirical proxies of some typical subprocesses of A such as physical vibrations and sounds of the right wave lengths (but fails to create empirical proxies for other typical subprocesses of a cat's purring, such as the relaxation of muscle tone).

human social behavior cannot take intentional states into account. The phenomenon we commonly call "anthropomorphizing" indicates, I think, that the performance conditions for social behavior operate with observable criteria that are schematic, involving generic Gestalts. As it appears, judging from our practices of interpretation, the upward position of the corners of a mouth does not need to be an *intentional* smile to count as a smile, nor does it need to resemble a human smile in all regards—a mouse can smile, and so can a sponge, a car, or a tree. That we use such general observable indicators of socially relevant actions and emotional states could be explained in evolutionary terms as follows. Surely it is preferable for humans to risk erroneous inclusions into the space of social interactions rather than erroneous exclusions; if we mistake something for a social agent, the error can be corrected without incurring social repercussions, but if we fail to recognize a social agent as such, this would amount to a social offense.

This concludes the distinctions in modes of simulations I wish to suggest here; further distinctions could be added if necessary. As the reader will have noticed, the definitions provided are not exact—they rely on a number of vague predicates ("functional," "observable," "typical"). In addition, they contain a context-relative parameter: the partition of a process. But, I submit, they capture five relevant intuitive notions of simulation: simulation as "a process with the same structure occurring in a different medium" (*functional replication*), simulation as "a process with similar structure that yields the same results as the original process" (*imitation*), simulation as "a process that yields the same observable results as the original process" (*mimicking*), simulation as "a process that yields the same relevant observable results as the original process" (*displaying*), and simulation as "a process that we can treat as observably similar to the original process" (*approximating*).

To illustrate how one might use these distinctions to clarify the capacities of robots, consider the robot *Leonardo*, created by Brian Scassellati (2002). Leonardo is programmed to produce, for a restricted interaction context, a representation of the beliefs and intentions of the person it is interacting with, to the extent that it can pass an empirical test for the capacity of "false-belief-attribution," the so-called "Sally-Ann test." In the philosophical discussion about sociality, false-belief-attribution counts as a prime example for social cognition. Given that Leonardo passes the Sally-Ann test, precisely what has been simulated: the Sally-Ann test, false-belief attribution, or even social cognition?

To bring the above definitions to bear on this question, assume that we operate with a partition P for the process A = *attributing-to-X-the-false-belief-that-p* that lists the process B = *passing-the-Sally-Ann-test* as one of the typical parts of A, and as another typical part the process C = *reacting-emotionally-to-the-belief-that-p*, belonging to the affective dimension of the theory of mind (cf. Kalbe et al., 2010). Then we should say that Leonardo does not replicate, imitate, mimick, or display, but merely approximates false-belief attribution A—if conceptualized as A_P—since Leonardo only realizes an empirical proxy for one of the processes that are typical for A. We cannot conclude from this that Leonardo thereby also approximates social cognition, since even if false-belief attribution were a typical part of social cognition, the empirical proxy of a typical part of X which is a typical part of Y is not an approximation of a typical part of Y. In contrast, assuming that we conceptualize B, the passing of the Sally-Ann test, in terms of a partition Q that lists as typical parts of B_Q the processes D = *uttering "p" in response to queries about X's belief*, E = *applying certain inference schemata to p*, F = *uttering "r" upon queries about an object's location in r*, and G = *tracing the locations of an object*, we can say that Leonardo mimicks the process of *passing-the-Sally-Ann-test*, since the robot realizes processes that are input-output equivalent to all the typical parts of B_Q.

In sum, the five definitions highlight that there are different degrees of simulation relative to structural and observable similarities. They also highlight that claims about one process X simulating another process Y depends on how we describe (partition) process Y. In the case of robotic simulations of human actions the following inverse relationship holds. If action A is described in terms of a coarse-grained partition that predominantly uses predicates for natural and functional process types (e.g., *uttering "p"* or *inferring*) a robot can be said to simulate action *A* to a high degree—it might be able to mimick, imitate, or even functionally replicate the action. *Vice versa*, if a fine-grained partition is used to describe action *A* and if the partition includes predicates for mental and reflectively normative process types (e.g., *judging* or *noticing*), a robot will likely achieve only low degrees of simulation—at least relative to the current state of technology, it will only be able to display or approximate actions under such a description. The five definitions of modes of simulations thus accommodate the description-dependency of claims about simulation while at the same time allowing us to formulate simulation claims more precisely.[20]

2.4 Simulatory Expansions of Interactions

The next step in developing a conceptual framework for simulated social interactions is to analyze the notion of interaction and relate the latter to the five modes of simulation as previously defined. Since there is no ready-made theory of concepts of interaction to draw on, let us set out with some preparatory reflections.[21]

The notion of interaction always signals a contrast to the mere co-occurrence of a plurality of processes. In the terminology introduced in the previous section, an interaction is a complex process as opposed to a collection of processes. However, in common-sense and scientific discourse we use the concept of an interaction in two senses. In its wide sense, the term denotes a complex *non-agentive occurrence* consisting of two or more non-agentive occurrences (often we use terms for objects or forces to refer to occurrences involved in an interaction). In contrast, in the term's narrow sense, "interaction" denotes a complex *action* consisting of two or more actions. For example, we use the wide sense when we describe corrosion (the formation of iron oxides) as an interaction between the physical processes that constitute iron atoms and those that constitute oxigene atoms, or define cell metabolism as an interaction between catabolic and anabolic processes. In contrast, a joint action that two or more agents intend to perform as such, such as playing football, discussing, carrying a table, or choire singing, is an interaction in the narrow, agentive, sense.

[20]The suggested classification of forms of simulation also should prove useful in the discussion of design issues in social robotics. For example, we can use it to compare and evaluate robotic simulations of an action A_P in terms of degrees of simulation; here one might refine *SIM-5* by introducing degrees of approximation; or it may be used to plan design goals, e.g., in order to decide whether there are any ethical reasons to aim for higher degrees of simulation than mimicking.

[21]Apart from Aristotle's conceptual analysis of interactions (cf. Gill, 2004), the most important source and resource for a future ontology of interactions is the work of Mark Bickhard, who combines empirical and conceptual research to promote "interactivism" both as a paradigm for empirical research and as a comprehensive theoretical stance or metaphysical view; cf. e.g. Bickhard (2009a, 2009b) and his chapter in this volume.

How do non-agentive interactions (wide sense) relate to agentive interactions (narrow sense)? This is largely unexplored territory in ontology. The question has been touched upon occasionally, e.g., when environmental ethicists consider whether the agents of a collection of actions X_n are responsible for the unintended interactive product of the processes that are, or result from, X_n (cf. e.g. Björnsson, 2011; Petersson, 2013). But currently we still lack a sufficiently comprehensive theory of multi-agent actions that would relate complex actions, collections of actions, complex non-agentive occurrences, and collections of non-agentive occurrences in all possible combinations.

In particular, we lack theoretical concepts for interactions that cannot be subsumed under either the wide or the narrow sense of interaction. Besides complex non-agentive occurrences whose parts are non-agentive occurrences (the first, wide sense), and complex actions whose parts are actions (the second, narrow sense), there are complex processes which have both actions and non-agentive occurrences as parts, and which may be either intended by an agent (third sense) or not intended (fourth sense).

The third sense of interaction is the primary target in our present context. In accordance with the four different senses of interaction identified, let us speak of *four kinds* of interaction—which each may be further differentiated into subordinate *types* of interaction—and let subscripts ("interaction$_1$" etc.) indicate which of the four kinds of interaction is at issue. We engage in an interaction$_3$ whenever we act in response to natural processes that are generated or shaped by our actions. For example, a basketball player doing solo dribbling drills engages herself in an interaction$_3$—the ball bounces in response to her actions and *vice versa*. Similarly, playing a single-player computer game can qualify as an interaction$_3$ (at least if we consider the events on the screen as natural processes rather than as the actions of the programmer). It is important to note that interactions$_3$ do not need to be initiated by the agent—consider a dynamic flood response, by building dams here and opening channels over there, performed dynamically in response to the spread and flow of the collecting waters. However, whether initiated or not, it is a hallmark of an interaction$_3$ that the agent intends to perform it by performing the actions that are among its parts. Interactions$_3$ thus resemble complex actions (interaction$_2$), and I shall call them "asymmetric" complex actions.

With these preliminary considerations in place, let us now turn to defining the two notions of interaction that are most relevant for present purposes.

Definition 7 (Interaction$_2$) A process system Σ is an *interaction$_2$* iff Σ is a complex action A_P and all parts of A_P are (simple, complex, or collective) actions B_i.

Definition 8 (Interaction$_3$) A process system Σ is an *interaction$_3$* iff Σ is a complex action A_P, some of the parts of A_P are (simple, complex, or collective) actions B_i, and some of the parts of A_P are non-agentive (simple, complex, or collective) occurrences C_i.[22]

Central for these two definitions is the distinction between actions and non-agentive occurrences, which is notoriously difficult to define. In the present context we can circumnavigate this difficulty by operating with the following simple condition. A process A_P is an action iff an agent can intend to do A_P.

To illustrate that this condition is sufficiently discriminative, consider a proto-typical interaction$_2$, a strategic board game such as chess. Each "move" of the game is a process A_P that one of the players intended to do (under the description A_P) and each player intends to participate in the game *as* an interaction$_2$, i.e., each player considers the processes executed by the other player as actions—as something that an agent can intend to do. In contrast, when two musicians perform a duo, each musician will react to sounds and phrases understood as non-agentive acoustic occurrences (one can intend to do what it takes to produce a certain sound on the instrument, but one cannot intend to sound); nevertheless each musician intends to bring about the interaction as complex process where the musician's actions respond to and influence sounds from another source. Thus a duo is the simultaneous occurrence of two interactions$_3$, generated by two musicians.

These illustrations show, first, that the suggested requirement for actions is subtle enough to dissociate the question "which parts of the interaction are actions?" from the question "which kinds of entities, humans or non-humans, cause the processes that are parts of the interaction?" Second, the examples also make plausible that interactions$_3$ are complex actions rather than complex occurrences—in this type of interaction the agent(s) who intend(s) to do the actions B_i that are part of A_P, also intend(s) to do A_P.

Precisely what is involved when we intend to do A_P has been much debated in the philosophy of action for the last half century it least. But even the least demanding conceptions of intentions take the latter to involve (at least) three abilities: (i) the ability to form beliefs about the process type A_P as a standard for action and interaction, i.e., as a template all parts of which *should* be realized if one wishes to realize A_P, possibly as an element within an interaction context, (ii) the ability to understand A_P as a process that underlies normative judgments (i.e., as an action that should, must, may etc. be done) and thus generates commitments, and (iii) the ability to initiate a process that is an instance of A_P and to pursue its realization despite

[22]Since the two definitions are to be as generic as possible, no requirements for dynamic, temporal, or spatial relationships among the parts of an interaction have been added. That is, I am assuming here that the processes that are the parts of an interaction may occur all or partly simultaneously, or with overlap, or sequentially in series. But we can easily introduce types and subtypes of interactions$_i$ by specifying which temporal, spatial, and dynamic relationships need to hold among the parts of the interaction in question. To simplify the exposition I omit here and in the following definitions specifications of the partition levels at which the parts of A_P are situated; as mentioned above, the embedding ontological framework of *General Process Theory* operates with a non-transitive part relation and parts are indexed to partition levels.

some changes in the action environment.[23] At the current state of technology robots lack at least the first two of these abilities; thus they cannot intend to do A_P, and thus they cannot realize an action. But they may realize non-agentive occurrences that simulate an action—currently at least in the sense of approximating, displaying, on mimicking.[24] Thus we can define the class of human-robot interactions as follows.

Definition 9 (Human-robot interaction) A process system Σ is a *human-robot interaction* iff

1. Σ has two process subsystems H and R which satisfy the predicates "human agent" and "robotic agent," respectively,
2. Σ is a complex action A_P and for any part X of A_P it holds that:

   a. X is either an action or the simulation of an action, and
   b. X is an action iff X is part of H, and
   c. X is the simulation of an action iff X is part of R.

This definition can be made more precise in a number of ways but the current version suffices, I trust, to convey the general idea that human-robot interactions are interactions$_3$.

In order to gain more specific descriptions of human-robot interactions we need to indicate the mode in which the robot simulates an action. For this purpose we first need to diversify the simulatory expansion of an interaction$_2$ as follows:

Definition 10 (Simulatory expansion matrix) For any process type A_P with n part processes, the *simulatory expansion matrix of* A_P is a $6^n \times n$ matrix of process types. Each row of this matrix consists of a sequence of n process types C_i so that each C_i is either identical with or a simulation of a process B_i just in case B_i is part of A_P.

To offer a toy example, let us assume that the process type A_P = *welcoming* has the following parts according to the partition P: $\langle B_1$ = *approaching target person*, B_2 = *establishing eye contact*, B_3 = *opening one's arms*, B_4 = *uttering in a friendly tone of voice "hello!"*$\rangle$. A section (!) of the simulatory expansion matrix for A_P = *welcoming* would then look as in Table 2.1.[25]

Each row in the simulatory expansion matrix of an action A_P describes one of the many ways of simulating A_P; that is, each row states a complex mode of simulation that can be formally represented as an ordered sequence of the modes in which each part of A_P is simulated. Using the expansion matrix of our simple example, Honda's ASIMO robot realizes simulatory expansion E_1, while H. Ishiguro's ROBOVIE robot, which is teleoperated and thus arguably *realizes* utterances, realizes the simulatory expansion E_2.

[23] For an overview over current accounts see e.g. Setiya (2007).

[24] To keep all options open, I will here also assume that it is conceptually possible to entertain the thesis that future robots may be able to imitate or functionally replicate *intending to do X*.

[25] Abbreviations for the modes of simulation are used as names for occurrences that simulate the action in the relevant column.

Table 2.1 A section of the simulatory expansion matrix for A_P = *welcoming*

	Approaching	Eye-contact	Open-arms	Utterance
⋮	⋮	⋮	⋮	⋮
E_1:	Imit	Dis	Mim	App
E_2:	Dis	Mim	App	Real
E_3:	Mim	Mim	Mim	Mim
E_4:	Dis	Dis	Mim	Dis
E_5:	Dis	App	App	App
⋮	⋮	⋮	⋮	⋮

Simulatory expansions can be defined for any sort of process, simple, complex, and collective, and both for actions and non-agentive occurrences. For complex processes the array of simulatory expansions quickly becomes very large. In general, for any kind of interaction $\mathfrak{I}$ with m agents and n parts, the simulatory expansion matrix for $\mathfrak{I}$ states all combinatorial possibilities for the realization of all parts of the interaction or for their simulation in any of the six modes, for all agents. For the case of an interaction of two agents the simulatory expansion matrix of this interaction has the dimensions $(6^n \times n) \times (6^n \times n)$.

Let us here focus on two-agent interactions$_2$ (i.e., complex actions) and let "$A_P[S1 : B1_1, \ldots, B1_k, S2 : B2_1, \ldots, B2_h]$" be shorthand for "the interaction$_2$ A_P whose parts are actions $B1_1, \ldots, B1_k$ performed by agent $S1$ and actions $B2_1, \ldots, B2_h$ performed by agent $S2$, as structured by partition P." The following definition introduces a special subtype of simulatory expansion matrices for interactions$_2$.

Definition 11 (asymmetric simulatory expansion matrix) The *asymmetric simulatory expansion matrix* of an interaction$_2$ $A_P[S1 : B1_1, \ldots, B1_k, S2 : B2_1, \ldots, B2_h]$ is a $k \times (6^h \times h)$ matrix which correlates each of the actions $B1_1, \ldots, B1_k$ of agent $S1$ with any of the possible $6^h - 1$ simulatory expansions of the actions $B2_1, \ldots, B2_h$ of agent $S2$.

In other words, the asymmetric simulatory expansion matrices of two-agent interactions$_2$ is a list of all interactions$_3$ that can be generated by simulatory expansions of all (and only) the actions of one of the two agents. Let us call any row in this matrix an *asymmetric simulatory expansion* of the relevant interaction$_2$.

2.5 Simulated Social Interactions

On the basis of the definitions presented in the two preceding sections we can reformulate the familiar *topos* of what robots can (or should) and can't (or shouldn't) do by asking "Which of the asymmetric simulatory expansions of one specific interaction$_2$ is a possible or desirable human-robot interaction?" This way of posing the question allows for differentiated answers.[26] More importantly, the core question of robot sociality can now be stated as follows:

(Q1) For any interaction$_2$ A_P that is a social interaction, which asymmetric simulatory expansions of A_P qualify as social interactions?

At first glance, it might seem that the answer to question (Q1) is quite straightforward: none. If we take a (again, for simplicity: two-agent) social interaction to be part of "social reality" and postulate, following Searle (2010, p. 8), that this requires "acceptance or recognition" by both agents for the "statuses" of the actions they perform, a social interaction implies that its parts are social actions. Thus, one might argue, since no simulation of an action can qualify as a social action, it follows that none of the asymmetric simulatory expansions of a social interaction is a social interaction; *a fortiori*, no human-robot interaction—which are asymmetric simulatory expansions of a social interaction at best—can qualify as a social interaction.

But matters are more complex. There are good reasons to question the philosophical standard model of social interactions, which still follows largely Hegelian lines and calls for joint "recognition" of normative statuses or for various forms of reflective collective intentionality (cf. e.g. Laitinen, 2011; Tuomela, 2013; Gilbert, 2008). In fact, as Cerulo (2009) has pointed out, the intentionalist notions of sociality in philosophy and the social sciences that exclude nonhumans are hardly ever argued for but presumed from the outset. Moreover, any principled exclusion can be challenged on empirical grounds in two ways. On the one hand, social robotics research itself provides strong evidence for reviewing the principled exclusion of nonhumans, since people prefer to use social vocabulary to describe their dealings with robots.[27] On the other hand, and more importantly, the intentionalist standard model of sociality does not sit well with recent research on the conditions of sociality in cognitive science, anthropology, psychology, and linguistics. For in view of this research it is far from clear that human social interactions are in all cases interactions$_2$.

As anthropologists and linguists have argued, sociality requires patterns of interactive practices of turn-taking organization and mutual orientation (cf. e.g. Schegloff, 2006; Goodwin, 2006), the elements of which are non-agentive occurrences—they are pre-conscious routines acquired by imitation learning. Similarly, cognitive scientists have identified pre-conscious "mechanisms" of social cognition, such as epistemic alignment (Samson, Apperly, Braithwaite, Andrews, & Bodley

[26] For example, surveying the set of asymmetric simulatory expansions of an interaction$_2$ one could investigate whether there are dependencies among expansions ("if this partial action is merely displayed, then that must be mimicked", etc.) or, *vice versa*, one could try to identify certain clusters of asymmetric simulatory expansions ("this sort of simulation succeeds for all interactions$_2$ that are short term/in educational contexts/involve touch etc."). Answers can be further tailored to technical possibility versus practical feasibility, or practical versus moral desirability.

[27] Cf. Turkle (2011). See Hakli (2014) for a discussion of possible methodological strategies of accommodating these phenomena for a theory of sociality.

Scott, 2010), sensory contagion (Blakemore, Bristow, Bird, Frith, & Ward, 2005), and "implicit" learning and tracking processes (Frith & Frith, 2012). Finally, from the side of evolutionary biology it has been pointed out that sociality can be explained without presupposing mindreading, i.e., the inferential capacities of a theory of mind, but rather on the basis of practices of conditioning (Mameli, 2001).

All these results can be taken to suggest that all human social interactions are $interactions_3$ consisting of a mixture of actions and non-agentive occurrences.[28] In short, recent empirical research suggests that there is much more involved in human social interaction than what we do or can intend.

Matters are further complicated by the fact that, as noted above, the definition and demarcation of social actions should take into account differences in point of view—differences between the participant's second person point of view and the observer's or third-person point of view.[29] Furthermore, we need to distinguish the *observer's* third person point of view from the point of view of the *omniscient* third person presupposed in philosophical analyses of the *concepts* of social interaction.[30] If we reflect on normative content assuming an omniscient third person point of view, it makes sense to postulate requirements of mutual reciprocal acknowledgements of intentionality, while from the viewpoint of the empirical observer, who does not have epistemic access to the intentional states of another agent, definitions of social interactions can only refer to the internal states of others in terms of manifestations of such states. Arguably, one cannot decide on (Q1) without first clarifying the more foundational methodological question (Q0):

(Q0) Which of these three viewpoints, two empirical and one normative, has the best claim to delivering the "right" account of sociality? Or should we consider sociality as a complex phenomenon that involves the facts of all three perspectives?

Whichever stance we take on (Q0), the theoretical tools of the "simulatory expansion matrix" of an action (in a given partition) as defined above should prove helpful for formulating the relevant claims more precisely, I submit. For example, compare

[28]To call such preconscious, non-intended occurrences "non-agentive" may be problematic—one might agree with my criterion for agency above that if an occurrence is an action then it must be possible to intend it, but deny that the implication also holds in the other direction; all processes that occur as parts of an intended action are agentive, one might say, even though they are not intended. I must bracket this issue here.

[29]Phenomenological analyses of forms of responsiveness long have drawn attention to the importance of this difference, but there is also increasing interest in "second person cognitive science," (cf. e.g. Reddy, 2008; Schilbach et al., 2013).

[30]Elsewhere (see Seibt 2018) I argue that for the purposes of attributing responsibility in a sufficiently differentiated fashion we need to distinguish between: (i) the *second person* point of view of the human interactor with the robot; (ii) the *internal third person* point of view of the roboticist who designs the interaction; (iii) the *external third person* point of view of the observer of a human-robot interaction; and (iv) the *omniscient third person* point of view of the cultural community evaluating the human-robot interaction and its effects on the surrounding context relative to the community's norms and values.

the following two characterizations of social interactions. The first one, (A), is the omniscient third person account of social interactions requiring joint knowledge about the mutual acknowledgement of norms of action (Gilbert, 2014); the second one, (B), is an account from a second person point of view in terms of the disposition to respond to the encounter of a conspecific in ways that express certain narrative-practical competences concerning action explanation (Hutto, 2012). If we choose characterization (A) as the "baseline" of normal social interaction, we could discuss (B) as an account of sociality that focuses only on those entries in the simulatory expansion matrices of an interaction A_P that realize simulation mode 4 (*displaying*) for all partial actions of A_P that are not bodily movements.

Once we have made headway on the question (Q0) and decided where to set the bar of the baseline of a "real" social interaction and from which viewpoint, we can then address the demarcation question (Q1). To illustrate which form such demarcation questions will take, assume we operate with baseline (A) as just stated (the "reciprocity account") and consider again the simplified example of an expansion matrix given above. Imagine I instantiate the simulatory expansion E_1 of *welcoming*; that is, I happen to walk precisely into your direction, thereby imitating the intentional action of *approaching you*, I happen to look approximately at you, thereby displaying the intentional action of *seeking eye contact*, I happen to stretch my arms, thereby mimicking the intentional action of *opening my arms*, and finally, as I am almost bumping into you, produce purely reflectively a "hello!", thereby approximating the intentional action of *uttering "hello!"*—have I been welcoming you? Or again, consider a simulatory expansion E_6, where all the parts of *welcoming* are realized apart from the last, the meaningful and friendly utterance of a welcome, which is only mimicked—should we count this, from a second or third-person perspective as a social action? E6 is the behavior of a person in the autism spectrum.

In sum, there are sufficient grounds to reject the idea that demarcation questions have straightforward answers. Which robotic occurrences are acceptable contributions to a social interaction depends not only on which forms of simulations we find acceptable, but also on the viewpoint we adopt for the definition of a social interaction in the first place—what appears as a simulation from one viewpoint (and relative to one account) is just the normal performance of sociality from another viewpoint (and relative to another account). The best strategy for coming to grips with the demarcation question may be to (i) adopt a maximally demanding baseline (i.e., to adopt the most restrictive realization requirements for social interactions from the omniscient third person point of view), (ii) to state the symmetric and asymmetric simulatory expansions for some typical social interactions, and then (iii) to discuss which regions of the expansion matrix should count as social interactions, drawing on alternative and less restrictive definitions of sociality.

2.6 Conclusion

The aim of this paper was to offer in outline a conceptual framework for the classification of forms of human-robot sociality. I have tried to differentiate between make-believe and fictional interactions ("treating as if"), social "taking as," and various forms of the "as if" of simulation. Since social terms are performative-ascriptive, I argued, they cannot be used in fictionalizing contexts, as social roboticists are wont to do when describing social robotics applications. Thus the difficult task of conceptualizing what happens in human-robot interactions, the "soft problem for the ontology of social robotics," cannot be addressed by a fictionalist "quick fix." In order to solve the "soft problem" we need to develop an ontology of simulated social interaction (SISI), I suggested, and sketched some possible trajectories for how one can approach this task. On the background of an embedding process ontology I have offered five definitions to distinguish varieties of the simulatory "as if" of human actions and interactions, in terms of five modes of simulations: *functionally replicating, imitating, mimicking, displaying*, and *approximating*.

According to sketch of SISI presented in this chapter, human-robot interactions are instantiations of asymmetric simulatory expansions of social interactions. The simple definitional tools of SISI allow us to develop classifications in terms of which we can formulate the questions of *robo-ontology* with greater precision. But the framework also holds out the promise, I believe, of facilitating the construction of a unified account of sociality.[31] For example, as I highlighted above, relative to standard "intentionalist" accounts of sociality in philosophy, also human-human interactions instantiate asymmetric simulatory expansions of social interactions, since not all of what we socially do is the result of conscious norm-guided actions. SISI can be used as a conceptual interface between philosophical and empirical research on human sociality—it allows us to investigate which simulatory expansions of social interactions in the standard intentionalist sense, i.e., using notions of sociality championed in philosophy, should be included in the scope of social interactions, given empirical research results on human sociality undertaken on the basis of non-intentionalist notions of sociality. As cognitive scientists, evolutionary biologists, anthropologists, and linguists have argued, the phenomenon of sociality does not hinge on the capacity of norm-guided action but begins much earlier, with joint attention to basic patterns: response rhythms and turn taking, and other elements of our pre-conscious capacity of imitating a praxis. Greater attention to the different modes and ways in which an intentional action can be simulated

[31]The conceptual tools of SISI are particularly basic (since it is grounded in a foundational ontology) yet precise and—due to its simple combinatorial strategies—highly expressive. It is therefore possible to translate into the classificatory framework of SISI other proposals of distinctions in capacities for moral agency (see e.g., Wallach & Allen, 2009), or in asymmetric forms of collective agency (see in particular the interesting contributions to Misselhorn, 2015); the details of these embeddings are yet to be worked out.

can facilitate reflections on how much (or how little) mentality we should require for a definition of social action. A fully worked out ontology of simulated social interactions might be the key to a comprehensive account of sociality where we can map out the relationships between empirical and conceptual analyses of sociality more clearly.

In short, then, the "soft problem" of robo-ontology, i.e., the problem of how to describe human interactions with "social" robots, motivates the development of an ontology of simulated social interactions and at once provides an important heuristics for an integrated account of sociality where empirical results and normative insights are combined.

Using the simple classificatory tools of SISI we receive new analytical perspectives on the ontology of the design process in robotics. As I argue elsewhere (Seibt, 2018), in order to generate sufficiently differentiated ascriptions of responsibility we can model design processes as new forms of collective intentionality, with more or less emphasis on engendering human-robot interactions in the so-called "we-mode" or the "I-mode."

It must be stressed, however, that the suggested approach at best yields an empirically grounded taxonomy of asymmetric simulated social interactions, for humans and robots. Such a taxonomy cannot enlighten us with respect to the "hard problem" of robo-ontology, namely, the question of what, if anything, is lost when a process is perfectly simulated (in the sense of functional replication as defined above). But discussion of taxonomic questions might help us to decide whether and where the hard problem matters at all.

References

Bernstein, D., & Crowley, K. (2008). Searching for signs of intelligent life: An investigation of young children's beliefs about robot intelligence. *The Journal of the Learning Sciences, 17*(2), 225–247.

Bickhard, M. H. (2009a). Interactivism: A manifesto. *New Ideas in Psychology, 27*, 85–89.

Bickhard, M. H. (2009b). The interactivist model. *Synthese, 166*, 547–591.

Björnsson, G. (2011). Joint responsibility with individual control. In *Moral Responsibility Beyond Free Will and Determinism* (Vol. 27, pp. 181–199). New York: Springer.

Blakemore, S.-J., Bristow, D., Bird, G., Frith, C., & Ward, J. (2005). Somatosensory activations during the observation of touch and a case of vision–touch synaesthesia. *Brain, 128*(7), 1571–1583.

Breazeal, C. (2002). *Designing Sociable Robots*. Cambridge, MA: MIT Press.

Breazeal, C. (2003). Towards sociable robots. *Robotics and Autonomous Systems, 42*, 167–175.

Cabibihan, J.-J., Javed, H., Ang, M. Jr., & Aljunied, S. M. (2013). Why robots? A survey on the roles and benefits of social robots in the therapy of children with autism. *International Journal of Social Robotics, 5*(4), 593–618.

Cerulo, K. A. (2009). Nonhumans in social interaction. *Annual Review of Sociology, 35*, 531–552.

Clark, H. H. (2008). Talking as if. In *3rd ACM/IEEE International Conference on Human-Robot Interaction (HRI), 2008* (pp. 393–393). Piscataway: IEEE.

Coeckelbergh, M. (2010). Health care, capabilities, and ai assistive technologies. *Ethical Theory and Moral Practice, 13*(2), 181–190.

Coeckelbergh, M. (2012). *Growing moral relations: Critique of moral status ascription*. Houndmills, Basingstoke / New York: Palgrave Macmillan.
Danish Ethical Council. (2010). *Sociale robotter. Udtalelse fra det etiske råd*. Retrieved from http://www.etiskraad.dk/~/media/Etisk-Raad/Etiske-Temaer/Optimering-af-mennesket/Publikationer/Udtalelse-om-sociale-robotter.pdf. (Accessed 13 Sep 2016)
Dautenhahn, K. (2014). Human-Robot Interaction. In M. Soegaard & R. F. Dam (Eds.), *The encyclopedia of human-computer interaction* (2nd ed.). Aarhus/Denmark: The Interaction Design Foundation. (Available online: https://www.interaction-design.org/encyclopedia/human-robotinteraction.html)
Enfield, N. J., & Levinson, S. C. (Eds.). (2006). *Roots of human sociality*. New York: Berg Publishers.
Fong, T., Nourbakhsh, I., & Dautenhahn, K. (2003). A Survey of socially interactive robots. *Robotics and Autonomous Systems, 42*, 143–166.
Frith, C., & Frith, U. (2012). Mechanisms of metacognition. *Annual Review of Psychology, 63*, 287–313.
Gilbert, M. (2008). Social convention revisited. *Topoi, 27*(1–2), 5–16.
Gilbert, M. (2014). *Joint commitment: How we make the social world*. Oxford: Oxford University Press.
Gill, M. L. (2004). Aristotle's distinction between change and activity. *Axiomathes, 14*(1), 3–22.
Goodwin, C. (2006). Human sociality as mutual orientation in a rich interactive environment: Multimodal utterances and pointing in aphasia. In N. J. Enfield & S. C. Levinson (Eds.), *Roots of human sociality: Culture, cognition and interaction* (pp. 97–125). Oxford/New York: Berg Publishers.
Gunkel, D. (2012). *The machine question*. Cambridge, MA: MIT Press.
Hakli, R. (2014). Social robots and social interaction. In J. Seibt, R. Hakli, & M. Nørskov (Eds.), *Sociable Robots and the Future of Social Relations: Proceedings of Robo-Philosophy 2014* (Vol. 273, pp. 105–115). Amsterdam: IOS Press.
Hutto, D. D. (2012). *Folk psychological narratives: The sociocultural basis of understanding reasons*. Cambridge, MA: MIT Press.
Kahn, P. H., Freier, N. G., Friedman, B., Severson, R. L., & Feldman, E. N. (2004). Social and moral relationships with robotic others? In *13th IEEE International Workshop on Robot and Human Interactive Communication, 2004 (ROMAN 2004)*. (pp. 545–550). Piscataway: IEEE.
Kahn, P. H., Friedman, B., Perez-Granados, D. R., & Freier, N. G. (2004). Robotic pets in the lives of preschool children. In *CHI'04 Extended Abstracts on Human Factors in Computing Systems* (pp. 1449–1452). New York: ACM Press.
Kalbe, E., Schlegel, M., Sack, A. T., Nowak, D. A., Dafotakis, M., Bangard, C., ... Kessler, J. (2010). Dissociating cognitive from affective theory of mind: a TMS study. *Cortex, 46*(6), 769–780.
Laitinen, A. (2011). Recognition, acknowledgement, and acceptance. In H. Ikäheimo & A. Laitinen (Eds.), *Recognition and social ontology* (pp. 309–348). Leiden/Boston: Brill.
Leyzberg, D., Avrunin, E., Liu, J., & Scassellati, B. (2011). Robots that express emotion elicit better human teaching. In *Proceedings of the 6th International Conference on Human-Robot Interaction* (pp. 347–354).
Mameli, M. (2001). Mindreading, mindshaping, and evolution. *Biology and Philosophy, 16*(5), 595–626.
Misselhorn, C. (Ed.). (2015). *Collective agency and cooperation in natural and artificial systems: Explanation, implementation and simulation* (Vol. 122). Cham: Springer.
Petersson, B. (2013). Co-responsibility and causal involvement. *Philosophia, 41*(3), 847–866.
Reddy, V. (2008). *How infants know minds*. Cambridge, MA: Harvard University Press.
Samson, D., Apperly, I. A., Braithwaite, J. J., Andrews, B. J., & Bodley Scott, S. E. (2010). Seeing it their way: evidence for rapid and involuntary computation of what other people see. *Journal of Experimental Psychology: Human Perception and Performance, 36*(5), 1255.
Scassellati, B. (2002). Theory of mind for a humanoid robot. *Autonomous Robots, 12*(1), 13–24.

Schegloff, E. A. (2006). Interaction: The infrastructure for social institutions, the natural ecological niche for language, and the arena in which culture is enacted. In N. J. Enfield & S. C. Levinson (Eds.), *Roots of human sociality: Culture, cognition and interaction* (pp. 70–96). Oxford/New York: Berg Publishers.

Schilbach, L., Timmermans, B., Reddy, V., Costall, A., Bente, G., Schlicht, T., & Vogeley, K. (2013). Toward a second-person neuroscience. *Behavioral and Brain Sciences, 36*(4), 393–414.

Searle, J. R. (2010). *Making the social world: The structure of human civilization*. Oxford: Oxford University Press.

Seibt, J. (2005). *General processes–A study in Ontological category construction*. Konstanz, Germany: Habilitations thesis at the University of Konstanz.

Seibt, J. (2009). Forms of emergent interaction in general process theory. *Synthese, 166*(3), 479–512.

Seibt, J. (2014a). Non-transitive parthood, leveled mereology and the representation of emergent parts of processes. *Grazer Philosophische Studien, 91*, 165–191.

Seibt, J. (2014b). Varieties of the 'as if': Five ways to simulate an action. In J. Seibt, R. Hakli, & M. Nørskov (Eds.), *Sociable robots and the future of social relations: Proceedings of Robo-Philosophy 2014* (Vol. 273, pp. 97–105). Amsterdam: IOS Press.

Seibt, J. (2018, forthcoming). The Ontology of Simulated Social Interaction–How to Attribute Sociality, Collective Agency, and Responsibility in Human-Robot Interaction. In J. Seibt, R. Hakli, & M. Nørskov (Eds.), *Robophilosophy: Philosophy of, for, and by social robotics*. MIT Press.

Setiya, K. (2007). *Reasons without rationalism*. Princeton: Princeton University Press.

Sharkey, A., & Sharkey, N. (2012). Granny and the robots: ethical issues in robot care for the elderly. *Ethics and Information Technology, 14*(1), 27–40.

Sparrow, L., & Sparrow, R. (2006). In the hands of machines? The future of aged care. *Minds and Machines, 16*(2), 141–161.

Sullins, J. P. (2008). Friends by design: A design philosophy for personal robotics technology. In P. Kroes, P. E. Vermaas, A. Light, & S. A. Moore (Eds.), *Philosophy and design: From engineering to architecture* (pp. 143–157). Dordrecht: Springer.

Tuomela, R. (2013). *Social ontology: Collective intentionality and group agents*. Oxford: Oxford University Press.

Turkle, S. (2011). *Alone together: Why we expect more from technology and less from each other*. New York: Basic Books.

Vallor, S. (2011). Carebots and caregivers: Sustaining the ethical ideal of care in the 21st century. *Philosophy & Technology, 24*, 251–268.

Veruggio, G. (2006). The EURON roboethics roadmap. In *6th IEEE-RAS International Conference on Humanoid Robots, 2006* (pp. 612–617).

Wallach, W., & Allen, C. (2009). *Moral machines: Teaching robots right from wrong*. Oxford: Oxford University Press.

Walton, K. L. (1990). *Mimesis as make-believe: On the foundations of the representational arts*. Cambridge, MA: Harvard University Press.

Walton, K. L. (2005). Metaphor and prop oriented make-believe. In M. E. Kalderon (Ed.), *Fictionalism in metaphysics* (pp. 65–87). Oxford: Oxford University Press.

Chapter 3
Robot Sociality: Genuine or Simulation?

Mark H. Bickhard

Abstract It is clear that people can interact with programs and robots in ways that appear to be, and can seem to participants to be, social. Asking the question of whether or not such interactions could be genuinely social requires examining the nature of sociality and further examining what requirements are involved for the participants in such interactions to co-constitutively engage in genuine social realities – to constitute genuine social agents. I will attempt to address both issues. A further question is "Why ask the question?" Isn't "sociality" like a program in that simulating the running of a program *is* the running of a program – so sufficiently simulated sociality is genuine sociality? What more could be relevant and why? As I will explain, there are at least two sorts of answers to the question of why the difference between genuine performance and simulation matters: (1) to better understand the ontology of sociality and thereby its potentialities and ways in which "merely" simulated sociality might fall short, especially of the developmental and historistic potentialities of sociality, and (2) to better understand the issues of ethics surrounding interactions among and between humans and robots.

Keywords Social ontology • Persons • Robot sociality • Interactivism

3.1 Introduction

This paper has three sequential issues: (1) The ontology of sociality, (2) Normative stakes in sociality, and (3) Robots and sociality. In the first discussion, I outline a model of social ontology, in preparation for the question of whether and in what sense robots might be capable of genuine participation in such an ontology. The second discussion focuses on an aspect of sociality that is, on the one hand, essential to genuine sociality, and, on the other hand, problematic for robots. The third

M.H. Bickhard (✉)
Department of Philosophy, Lehigh University, 15 University Drive, 18015, Bethlehem, PA, USA

Department of Psychology, Lehigh University, 17 Memorial Drive East, 18015, Bethlehem, PA, USA
e-mail: mhb0@lehigh.edu; www.bickhard.ws

R. Hakli, J. Seibt (eds.), *Sociality and Normativity for Robots*, Studies in the Philosophy of Sociality 9, DOI 10.1007/978-3-319-53133-5_3

section elaborates this problematic characteristic, and addresses whether it might be overcome, and some of its implications for the possibilities of robots as moral agents.

These three focal discussions are developed within a framework of a broader model that addresses the emergence of normative function, representation, cognition and language, social ontology, and the ontology of personhood. I will begin with an introduction to that framework.

3.2 Normative Emergence

3.2.1 Process and Emergence

A core point of this discussion is that human sociality involves forms of normativity that robots, in their current form, cannot capture, though they might be able to simulate that normativity up to a point. To explicate this point requires developing a model of normativity, especially normativity of representation and of motivation, and, in turn, of the emergence of such forms of normativity.

Emergence, however, is itself a problematic notion, so I begin with an argument that emergence is in fact not possible within a classical substance-metaphysical framework that we have inherited from the Greeks, but that it is possible, even ubiquitous, if we return "back to Heraclitus" and work within a process metaphysics. A process metaphysics not only, so I argue, makes genuine metaphysical emergence possible, but also overcomes some serious, if not fatal, problems with more standard substance or particle-based metaphysics.

I begin with Parmenides. Parmenides argued that change cannot occur (R. J. Campbell, 1992; Gill, 1989; Guthrie, 1965). This is at times interpreted as a response to Heraclitus' process model (Graham, 2006), though, whether or not that is correct, it certainly stands opposed to Heraclitus. The argument is that, when considering fundamental metaphysical reality, for something fundamental *A* to change into something also fundamental *B* would require that *A* disappear into nothingness and *B* appear – emerge – out of nothingness. "Nothingness" cannot exist: you cannot point to it (R. J. Campbell, 1992); "nothingness" here does not mean something like a vacuum (though that was a considered interpretation at times), but, rather, something itself fundamental, perhaps like a contemporary person trying to think about the "nothing" that "exists" outside of the universe. This argument has had an enormous influence on Western thought, in significant part because Plato and Aristotle gave it great credence.

Democritus and Empedocles both attempted to show how the appearance of change could occur without violating the Parmenidean argument. The key is to recognize that change does not occur at a fundamental level of metaphysics, but, rather, at a more "surface" level of appearances. For Empedocles, the fundamental metaphysics consisted of four substances, earth, air, fire, and water, and these did not change. What could change, however, were mixtures of these. The combinations of earth, air, fire, and water could vary even though the four substances themselves did not change. Democritus, with respect to this issue, had a similar model: the configurations of atoms could change, but the atoms themselves did not.

Aristotle had a sophisticated elaboration of such a model. For Aristotle, earth, air, fire, and water could change into each other, but they were not fundamental. He maintained the constraint that change could only occur with respect to a (more) fundamental unchanging substrate for change, either in the form of Prime Matter that never changes, or in a still more sophisticated version that turned on unchanging substrates of hot or cold or wet or dry (Gill, 1989).

In any case, this is the basic constraint that has come down to us as a dominant background assumption: change occurs only with an unchanging substrate. In contemporary views, this level of substrate is usually assumed to consist of basic particles.

Adopting such a framework has several consequences. Here are three that are of importance for the following discussion:

1. Stasis, lack of change, is the default condition. Change requires explanation.
2. Emergence is impossible. You cannot, for example, get a new substance out of Empedoclean earth, air, fire and water. Emergence, in fact, is precisely what these were intended to preclude – they (claim to) account for change *without* having to posit emergence.
3. A "natural" world is limned consisting of substances or particles, "causal" relations among them, factual relations among them (e.g., configurations of particles). This realm is split from a realm of normativity and intentionality – a roughly "mental" realm that does not seem to fit into, or together with, the "natural" realm.

Given this split, there are only three general possibilities: (1) Assume dual metaphysical realms, such as Aristotle's matter and form or the Cartesian dual substances. (2) Try to account for the "natural" world in terms of the "mental" world, as in some versions of idealism. (3) Try to account for the "mental" world in terms of the "natural" world. This latter is the dominant position today, in which naturalism is assumed, and is assumed to consist of some sort of physicalism.

We would, perhaps, like to account for the normative, intentional world as *emergent* within the natural world, but emergence is precluded by precisely the substance/particle assumptions that create the split in the first place. Adopting a process metaphysics, however, overturns all three of these consequences:

1. Change becomes the default, and constancy or stability requires explanation.
2. Emergence becomes possible in organizations of processes.[1]

[1]For discussions of Hume's and Kim's arguments that preclude emergence, explicit for Kim and implicit for Hume (see Bickhard, 2009b, R. J. Campbell & Bickhard, 2011 and below). I argue that both presuppose a particle or entity metaphysics.

3. With barriers to emergence removed, it becomes possible to explore models of the emergence of normativity and intentionality – the possibility of (re-) integrating the two realms.

3.2.2 *Why Process?*

From the perspective of a substance or entity or particle metaphysics, organization is not a legitimate locus of causality. Organization is not a "thing", and only things can have basic causality. So any apparent causality that is manifest by some new organization of, say, particles, is just the particle causality working its way out starting from whatever configuration or organization they started in. Organization is just "initial conditions" and has no possible causality itself. But if emergence is supposed to account for new causal powers in new organization, then a particle framework makes this impossible by assumption.

Processes, however, are intrinsically organized, and have whatever consequences they have for the rest of the world – whatever causal influences they have – partly and necessarily in terms of their organization. Processes are intrinsically organized, and that organization is not just stage setting for their ongoing flow. Within a process metaphysics, (causal) influence on the world is necessarily dependent on (among other things) organization; organization cannot be excluded as a legitimate locus for causality without removing causality from the universe.

New or different process organizations, thus, can yield new or different consequences for the world. They can yield emergent forms of influences on the world.

So, process enables emergence. Are there other reasons to accept a process metaphysics? Yes.

First, in a pure particle metaphysics, the probability of one dimensionless particle hitting another are zero. In such a universe, nothing would ever happen.

A common view today is that the world is constituted as point particles that interact with each other not by contact, but via fields. This is strictly not correct (see below), but it already forces the acceptance of organization as potential locus of "causal" power: fields are organized, and have whatever influence on the world that they do have in part in terms of those organizations.

Second, our best contemporary physics tells us that there are no particles (Aitchison, 1985; Aitchison & Hey, 1989; Bickhard, 2003; Brown & Harre, 1988; Cao, 1999; Davies, 1984; Halvorson & Clifton, 2002; Huggett, 2000; Kuhlmann, Lyre & Wayne, 2002; Sciama, 1991; Weinberg, 1977, 1995). Everything is quantum fields, and what remains of the notion of particle is only that quantum field interactions are intrinsically quantized. This quantization is of either whole integer or half integer multiples, just as is the waveforms in a guitar string: if held at two points, the guitar string vibrates with whole integer numbers of waves, while, if held at only one point, it oscillates with half integer multiples (consider, for example, the half integer wavelengths in a rope that is waved from one end). This is quantization, and there are no guitar sound particles; nor are there physical particles.

It might be pointed out that our best theories in physics might nevertheless not be correct. Yes, and, in fact, we have good reason to think that they are false as they currently stand. Nevertheless, they do support a process metaphysics, not a particle metaphysics, and, perhaps more importantly, we have empirical evidence that makes it essentially impossible to ever return to a particle framework, no matter what changes are made in the theories. For example, in a vacuum, according to quantum field theory, there is constant oscillatory activity, with wave-like processes emerging and disappearing – a kind of froth or foam of activity. If two very smooth metal plates are held (very) close to each other, they act similarly to the holding of a guitar string at two places: only whole wavelength activity can occur between the plates. In this manner, the activity between the plates is reduced relative to the activity outside of the plates, and this induces a pressure pushing the plates toward each other. This effect has been detected and measured, completely in accord with theoretical predictions, and there are no particles in this story (Sciama, 1991). In effect, the localities and isolations of a particle based assumption have been shown to be false, and, in that sense, there is no possibility of return to a particle metaphysics.

So, a process metaphysics enables emergence, potentially enabling models of the emergence of normativity and intentionality (and related phenomena), and it is the only kind of metaphysics that is viable given contemporary physics, both theoretical and empirical.

3.2.3 Maybe Causal Emergence, but Normative Emergence?

Shifting away from substance models to process models is a basic theme in the history of science. We have given up models of fire as the substance phlogiston, heat as caloric, magnetism as magnetic fluid, life as vital fluid, all in favor of process models of combustion, kinetic energy, field models, and various kinds of self-organizing, self-maintaining, processes. Normativity and intentionality, however, seem to pose special problems. They are not only the core of what was split off by the ancient delimitation of a substance world, they are also the kind of phenomena that Hume supposedly showed could not be derived from fact: no norms from facts; no ought from is.

Hume did not actually develop an argument, and there have been multiple attempts to fill in what that argument might be (Schurz, 1997). Nevertheless, his conclusion has had powerful influence. Hume's "argument" consists of the phrase "what seems altogether inconceivable":

> I cannot forbear adding to these reasonings an observation, which may, perhaps, be found of some importance. In every system of morality, which I have hitherto met with, I have always remark'd, that the author proceeds for some time in the ordinary way of reasoning, and establishes the being of a God, or makes observations concerning human affairs; when of a sudden I am surpriz'd to find, that instead of the usual copulations of propositions, *is*, and *is not*, I meet with no proposition that is not connected with an *ought*, or an *ought not*.

> This change is imperceptible; but is, however, of the last consequence. For as this *ought*, or *ought not*, expresses some new relation or affirmation, 'tis necessary that it shou'd be observ'd and explain'd; and at the same time that a reason should be given, for **what seems altogether inconceivable**, how this new relation can be a deduction from others, which are entirely different from it. But as authors do not commonly use this precaution, I shall presume to recommend it to the readers; and am persuaded, that this small attention wou'd subvert all the vulgar systems of morality, and let us see, that the distinction of vice and virtue is not founded merely on the relations of objects, nor is perceiv'd by reason. (Hume, 1978, Book III. Part I. Section I. 469–470, emphasis supplied)

I offer here an interpretation that makes Hume's point valid, though, so I will argue, unsound. Consider a valid conclusion based on premises that, by assumption, contain only factual terms. If that conclusion contains new terms, not present in the premises, how could they have been introduced? By definitions. The only form of definition that Hume knew about, a form descending from Aristotle, was what might be called abbreviatory definition, in which a term or phrase is defined as being equivalent to a defining term or phrase. What is crucial for current purposes about such definitions is that they can be "back translated" through. That is, for all instances of a defined term or phrase, the defining term or phrase can be substituted. Perhaps there will still be new terms, but these too can be eliminated in favor of the defining terms or phrases, and so on until all remaining terms in the conclusion are those that were in the original premises. These, by assumption, are all factual, so no non-factual – e.g., normative – terms can be validly introduced in the conclusions.

Notice, that although directed at deriving "ought" from "is", the general form of this argument is that nothing beyond new grammatical (or logical) configurations of premise terms can be validly deduced in conclusions. This is reminiscent of a basic particle point from Democritus: only new configurations, not new particles. In this generality, Hume's "argument" is an argument against any kind of emergence.

But it is unsound: it involves a false premise. The false premise is the assumption that all legitimate forms of definition permit back translation. Hume knew of no other kinds, but, since Hilbert, we do know of other kinds of definitions today. In particular, implicit definition (as distinguished from explicit definition, which *does* permit back translation) does not permit back translation. Implicit definition was introduced by Hilbert (and other authors around the turn of the 20th century) with respect to formal axiom systems. Consider an axiom of geometry: two points determine a line. In classical views, the terms in this axiom would have to themselves be further defined in terms of, among other things, what a point and a line are. These ultimately would be taken to denote things in the real world, or in some kind of formal world. For Hilbert, their interpretations were left free. Any interpretations that satisfied the relations within and among the axioms were legitimate; any such interpretation (using Tarskian language) constitutes a *model* of the axioms. So, for example, two points determine a line in the sense of the line that runs between the points, but also two lines determine a point – reversing the interpretation – in the sense of determining the point of intersection of the lines (so long as points at infinity are accepted for parallel lines): the terms "point" and "line" can be interpreted either way.

A set of axioms *implicitly defines* its class of models. What is important for my current purposes is that such an implicit definition is, on the one hand, completely legitimate, and forms part of the basis for contemporary model theory (e.g., Chang & Keisler, 1990), but, on the other hand, cannot be back translated through. If the above construction of what Hume had in mind is correct,[2] then his conclusion is unsound because it makes a false assumption about the nature of definition.[3]

If we shift to a process metaphysics, then, emergence becomes possible, and if we recognize that Hume's argument is based on a false assumption about the nature of definition (as well as mirroring a basic particle metaphysics consequence), then perhaps normative emergence becomes possible.

3.3 The Emergence of Normative Function

Normativity and intentionality pose particular problems for attempts to model their emergence. Physics in general provides multifold distinctions among and within a large class of variables – location, velocity, charge, etc. – but there is no sense in which there is an asymmetry among those distinctions that the normative asymmetry between, for example, functional and dysfunctional or true and false or good and bad might be based. There is, however, one basic exception to this point: thermodynamics. And it is a thermodynamic distinction that forms the ground for the model of normative functional emergence.

Within a process metaphysics, change is the default, but there are many ways in which *organizations* of process can manifest persistences and stabilities. One major class of such stabilities is constituted by what can be called *energy well* stabilities. These are organizations of processes that remain persistent unless and until some above threshold energy impinges on them. An atom is a canonical example: it is a furious process among quarks, gluons, and electrons, but it remains stable in its basic organization for potentially cosmic time periods.

Another class of stabilities is at the opposite extreme of thermodynamics. An atom, or a rock, can be isolated from the world, and it will remain stable. It will go to thermodynamic equilibrium and stay there. A candle flame, on the other hand, cannot be so isolated: if isolated it will go to equilibrium and *cease to exist*. A candle flame is – ontologically – far from thermodynamic equilibrium, and it cannot be isolated without disrupting the maintenance of that far-from-equilibrium condition. Energy well stabilities persist at equilibrium, and do not require maintenance; far-from-equilibrium processes can persist, but only if they

[2]If it is argued that it is not correct, then the burden is on whoever offers such a claim to make good on what Hume's argument could possibly have been (Schurz, 1997). The abbreviatory definition interpretation fits what Hume knew about definition, and does support his "argument" as valid.

[3]There are also non-formal senses of implicit definition (e.g. Bickhard, 2009b; Hale & Wright, 2000).

are maintained at far-from-equilibrium conditions. Far-from-equilibrium processes require maintenance in order to persist, and energy well stabilities do not. This is the basic asymmetry upon which normativity is based: maintenance of the far from equilibrium conditions essential for a far-from-equilibrium process is contributory – *functional* – for the persistent existence of that process.

A candle flame also illustrates another important property: *self*-maintenance. A candle flame contributes to its own stability; it tends to maintain the conditions for its own persistence: It maintains above combustion threshold temperatures, induces convection which brings in oxygen and removes waste products, and so on. Far-from-thermodynamic-equilibrium conditions are essential for all self-*organizing* processes, but candle flames illustrate this further property of self-*maintenance*.

A candle flame is self-maintaining, but if, for example, it runs out of wax, it has no way of detecting that nor of doing anything about it if it were to detect it. A bacterium, however, illustrates a further crucial property: that of *recursive* self-maintenance (Bickhard, 2009b). If a bacterium is swimming up a sugar gradient, it will tend to keep swimming – toward more sugar. If it detects going down a sugar gradient – away from sugar – it will tend to tumble for a moment and then take up swimming again. In these circumstances, swimming is contributory, functional, for the bacterium so long as it is swimming toward more sugar. But it is dysfunctional if it is swimming toward less sugar. The bacterium is capable of detecting something about its circumstances in the world and adopting activities that are functional within those circumstances (D. T. Campbell, 1974). It is able to maintain its condition of being self maintenant across changing conditions: it maintains self-maintenance – recursive self-maintenance.

3.3.1 Etiological Models of Normative Function

The dominant models of normative function today are etiological models (Millikan, 1984, 1993). The central idea of these models is that some organ in the body *has a function* if it has the right kind of etiology, generally an evolutionary or a learning etiology. I will focus on Millikan's model, in which the required history for having a function is an evolutionary history: ancestral organs have been selected during evolution for having the consequences that constitute the function of the kind of organ involved – e.g., the kidney has the function of filtering blood because ancestral kidneys were selected for filtering blood.[4]

There are several questions and problems with such models. I will focus on a couple of them. One central problem is that having a function is *constituted* in

[4]Note that, insofar as such models were to succeed in modeling normative function, they would violate Hume's argument. I have argued that Hume's argument is unsound, but the point that either something is wrong with Hume's argument or else these models cannot be correct is generally overlooked or ignored.

having the right (evolutionary) history. But differing etiologies, evolutionary or otherwise, can yield the same organization of processes. That is, differing histories can yield dynamically, causally, identical systems. The problem that this produces is illustrated by a thought experiment that Millikan discusses: suppose that the molecules in the air come together for form a lion in the corner of the room that, by assumption, is molecule by molecule identical to a lion in the zoo. The lion in the zoo has organs with the right evolutionary histories to have functions, but the lion in the corner, even though identical in a dynamic sense to the lion in the zoo, has no relevant history, and, thus, its organs have no functions. In this case, we have two lions that are causally identical, but one has functions and the other does not: function, in this model, is causally epiphenomenal.

A related problem is that the evolutionary history has to have a beginning, and issues become problematic when such beginnings are considered. Consider the first time that some organ, or part or location, in an organism is selected for having some consequence. After some number of generations of this (how many is not clear) descendent organs will have that consequence as a function. It might be considered that such selection is strictly causal, with no normative properties at all, till some sufficient number of generations has passed, at which point normativity emerges. But being selected for something involves that something, that consequence, being positive for the organism, contributing to the organism's existence, being *functional for* the organism in the sense of the thermodynamics based model outline above. Whether or not this model of function is accepted, it is difficult to avoid the conclusion that the basic selections, no matter how many or few the number of generations involved, is already a normative kind of phenomenon. It is already normative in the sense of being useful to the organism, though, perhaps, not yet in the sense of being designed for serving that function (Christensen & Bickhard, 2002).

One basic difference between etiological models and the thermodynamic interactive model is that etiological models focus on what it is to have a function, which lends itself to a kind of evolutionary (or learning) design sense of what is involved, while the interactive model focuses on what it is to serve a function, to be useful, which is most primitively based on thermodynamic considerations. If having a function is central, then serving a function must be addressed, and that is generally addressed with the notion that something serves a function when it "succeeds" in having the consequence that is the function that it has. So, an organ serves a function when it does what its ancestral organs were selected to do. This makes it difficult, for example, to account for "accidental", not "designed", usefulnesses, such as the function of leg muscles in aiding blood flow on long airplane flights.

If *serving* a function is central, then the notion of *having* a function must be explicated. In general, the idea here is that various activities in an organism *presuppose* activities of other parts of the organism in the sense that the given activities can succeed in serving their functions so long as the presupposed activities and consequences are available. Such functional presuppositions that such and such a consequence will be realized in such and such a location – e.g., blood filtered in kidney locations – is what constitutes *having* a function in this model. In this model, having a function is derivative from serving a function. There is no difficulty in this model in accounting for non-designed serving a function – serving a function that no organ *has*.[5]

3.4 Representation and Cognition

The notion of functional presupposition is crucial to a next major emergence: that of *representational truth value*. In general, when a living being engages in interactions with its environment, those interactions will be functional under some circumstances, but not functional under other circumstances. If an interaction is engaged in the "right" circumstances, it will succeed in serving whatever function(s) it might be associated with. Engaging in such an interaction, then, functionally presupposes that such supporting conditions hold, in the sense that engaging in such an interaction would not be functional if those supporting conditions did not hold, and the interaction in that case would not be true to its functional nature (R. J. Campbell, 2011): for the bacterium, swimming presupposes that its orientation is toward more sugar. This is an interaction level, environmental version of the functional presuppositions that hold among various organs within an organism that constitute those organs having functions.

Crucially, if an interaction is engaged in and supporting circumstances hold, then the presupposition of those circumstances is correct – is *true*. The interaction is true to its functional nature if its functional presuppositions are true (R. J. Campbell, 2011). If the bacterium is oriented up a sugar gradient, then the presupposition of the activity of swimming holds, and swimming will in fact be functional; it could be *dys*functional if the bacterium were swimming up a saccharin gradient, in which case the presupposition would be false. Here we have the emergence of representational truth value – the basic normativity of representation – within the framework of the functionality of various kinds of interactions between the system and its environment.

This is a primitive version of representation, but the central model has resources to account for more complex forms of representation. Consider a frog that might have a possibility of flicking its tongue in one direction to eat a fly, another direction to eat a different fly, and downward to eat a worm. The frog must have some functional way in which to indicate these possibilities, and some way of selecting among them. For my current purposes, the central point is that indications of interactive possibilities can branch into multiple possibilities.

They can also conditionally iterate: the frog, to continue that example, might also have the possibility of shifting to its left, at which point a different worm might become accessible – some interactions can create the possibilities for other interactions. Such branching and iterating indications of interactive potentialities

[5]For further discussion and comparison, see (Christensen & Bickhard, 2002; Bickhard, 2009b).

can, in more complex organisms, form vast webs of indications, and it is within such webs that more complex forms of representation can emerge.

Consider a child's toy wooden block. There are many interactions possible with such a block, including multiple visual scans and manipulations, as well as throwing it, leaving it on the floor, and so on. There is a sub-web of such possibilities that has certain special properties: every scan or manipulation is reachable from every other scan or manipulation, with appropriate intermediary manipulations, and this internally completely reachable sub-web is itself invariant under various other things that can be done or could happen, such as throwing, leaving on the floor, etc. Some things, however will destroy the opportunities of the web, such as crushing or burning the block. In this manner, representations of, for example, small objects can be constructed. This is essentially a borrowing of Piaget's model of the representation of small objects into the interactive model.[6] Such borrowing is possible because both Piaget's and this model are action based.[7]

A complex agent, thus, must characterize its environment in terms of what interactions with that environment are possible (Bickhard, 2009b).[8] An agent must have such a functional characterization in order to be able to select among those possibilities in accordance with criteria such as goals – e.g., a goal of eating if the internal state is one of hunger (e.g., low blood sugar).[9] Such possibilities can be complex, and are constantly changing due to environmental processes as well as activities of the agent itself, so organizations – webs – of such *anticipations* of possibilities must be constantly maintained and updated. I call such complex webs the organism's *situation knowledge* – knowledge of what potential interactions are available in the current environment – and the maintenance and updating of situation knowledge is called *apperception*.[10]

---

[6]As in Piaget (1954). There is a substantial literature that claims that such knowledge is innate, and, thus, not really constructed. But there is also a large family of reasons to consider such claims to be based on incorrect theory and systemically faulty methodology (Allen & Bickhard, 2013).

[7]For discussion of these resources in more detail, including resources for representations of abstractions such as the number three, see, e.g., Bickhard (2009b).

[8]It should be noted that this model of representation is a model of functional emergence, not of consciousness or conscious representation. Neither the bacterium nor the frog, for example, have reflective consciousness. The overall model, however, does have some strong implications regarding consciousness, and entails: (1) that consciousness is not the unitary phenomenon that is commonly assumed, (2) that *some* properties of consciousness are in fact emergent in fairly simple organisms, and (3) that much of the mystery of how to naturalize consciousness evaporates when it is realized that mystery is a product of underlying false assumptions being made in the literature (Bickhard, 2005).

[9]Processes of selecting among trajectories of indicated interaction possibilities constitute the domain of *motivation*.

[10]These anticipatory webs have some similarities with Gibson's notion of affordance, though Gibson's affordances cannot have the kind of structure and organization mentioned above. For discussion of Gibson, see Bickhard & Richie (1983).

3.4.1 Encoding Models of Representation

Extant models of representation are versions, sometimes ingenious versions, of what can be called *encodingism*: the assumption that all representation is some form of encoding. There are multiple versions of encoding models, but they have in common an assumption that there is some special correspondence between representation and represented that constitutes a representational relationship. This special relationship is variously modeled as being causal, informational, structural, etiological, and so on. One further common characteristic is that such models assume that the special representational relationship is some version of looking backwards down the input stream to try to "represent" where that input stream is coming from–spectator models, in the language of Dewey (1960; Tiles, 1990).

Encodings certainly exist, witness Morse code, in which, for example, "···" encodes "s". Such codes can be useful: dots (and dashes) can be sent over telegraph wires while characters, such as "s", cannot. For another example, most of us would starve if computer codes were suddenly abolished: they are very useful in contemporary society. Furthermore, codes do not have to be conventional like Morse code or computer codes: Consider a neutrino count in a gold mine in North Dakota that encodes properties of fusion processes in the sun.

What all genuine codes have in common, however, is that in order for the code to exist, both ends of the encoding relationship must be known, and the relationship between the ends must be known: both "···" and "s", both the neutrino count and the fusion properties, and the relationships between them. Codes change the form of representation, and give us access to doing things with those encoding representations, such as inferring fusion properties or transmitting a message over a wire, that would otherwise be not possible or much more difficult. Crucially, codes require that the encoded end of the relationship *be already known*. Encountering "···" per se cannot give you any sense of "s", nor neutrino counts per se anything about fusion properties. Encodings cannot constitute representation of anything not already known about.

3.4.2 Problems with Encodingism

There is, in fact, a large family of fatal problems with encoding models, some known for millennia, some of more recent provenance. But encoding assumptions still dominate because there has not been an alternative on offer. The interactive model outlined above belongs in the tradition of pragmatism, which is historically relatively recent, and avoids the problems of spectator models in virtue of being future oriented, not past oriented.

One problem with encoding models is that they cannot account for the emergence of representation. You have to already have a representation of "s" in order to define "···" in terms of it. But what constitutes the representation of "s", and where did it come from? Encoding definitions can iterate, some things defined in terms of others, which are, in turn, defined in terms of others, but this must halt at some level of basic representations out of which all others must be defined. Fodor argues that this base must be innate (Fodor, 1981). But if representation cannot emerge, then evolution cannot generate representation any more than can learning or development. Conversely, if evolution *can* generate emergent representation, then Fodor offers no argument about why learning and development could not do the same. But representation did not exist 13 billions years ago and it does now: it has to have emerged. In fact, Fodor at times acknowledges something like this:

> I am inclined to think that the argument has to be wrong, that *a nativism pushed to that point becomes unsupportable, that something important must have been left aside.* What I think it shows is really not so much an a priori argument for nativism as that *there must be some notion of learning that is so incredibly different from the one we have imagined* that we don't even know what it would be like as things now stand.[11] (Piattelli-Palmarini, 1980, p. 269)

Another problem with encoding models is that they cannot account for organism-detectable representational error. In fact, they tend to have difficulties accounting for the very possibility of error at all, and *organism-detectable* error is not addressed. Attempts at modeling the "mere" possibility of error are from the perspective of some external observer of the organism, who is supposed to determine whether the organism's representations are correct or incorrect about something being represented in the environment – e.g., I think I see a cow, but my "external observer" can tell that it's really a horse on a dark night: in what sense is my representation of "cow" wrong and how can the external observer determine that?

But, if organism detectable error cannot occur, then error guided behavior and learning are not possible. We know that error guided behavior and learning occur, so any model that cannot account for organism-detectable error is thereby refuted. In general, the problem is that the only way I have to check my representation to see if it is correct is to use that same representation again. This is circular, and does not provide any knowledge of error.[12] Every standard model in the literature fails this criterion: Fodor, Millikan, Dretske, Cummins, symbol systems, connectionist models, and so on (Bickhard, 2009b, 2014).

---

[11] This is one of multiple places in which Fodor acknowledges that there are serious problems with his model. See, for another example, Fodor (1990).

[12] That this problem is equivalent to the classic radical skeptical argument provides some sense of its difficulty. The radical skeptical argument points out that, in order to check my representation, I would have to somehow step outside of myself and gain independent epistemic access to what I am trying to represent – become my own external observer – to be able to compare my representation with what is being represented. I cannot step outside of myself, so I cannot check my own representations. One common intuition is that I can check *consequences* of my representations: walk up closer to the "cow" and discover that it is in fact a horse. I think there is a germ of a correct approach here (anticipation), but, as stated, it simply checks one representation with another, and gives no reason to accept that either one is true or false. Such considerations can lead to pure coherence models. For further discussion, see Bickhard (in preparation).

The interactive model solves or dissolves both of these problems. Representation is *emergent* in particular kinds of functional processes (indications of interaction possibilities), and so does not require some base of already available representational atoms. And the possibility of *organism-detectable* error emerges in virtue of the future orientation of interactive anticipation: if the indicated interaction is engaged in and it does not proceed as anticipated, then the anticipation is falsified. Error detection is constituted not as peering back in time down the input stream, but in checking whether or not the future turns out to be as anticipated.

3.5 The Ontology of Sociality

Something interesting emerges when two or more complex agents are in each other's interactive presence: Complex agents pose special problems for each other, and solutions to those problems constitute social realities.[13] Sociality, then, – social ontology – arises intrinsically among complex agents (assuming that they arrive at solutions to the intrinsic problems).

3.5.1 Two Agents

Consider now two complex agents in each other's presence. Each seeks to interactively characterize their environments, which includes interactively characterizing the other agent. But the other agent's interactive potentialities are constituted in part by their own characterization of their environment, which includes the first agent. Each agent, then, needs to interactively characterize the other, including in terms of that agent's characterization of itself – which is constituted (in significant part) by the characterization of the other. There is regress here.

Each agent has an interest in resolving this characterization problem, but they cannot do so without arriving at compatible mutual characterizations of the situation together with each other's presence in that situation. This constitutes a coordination problem in the sense of Schelling (1963), and, thus, a resolution of the problem is (roughly) a convention in the sense of Lewis (1969).[14] I argue that social realities – *all* (forms of) social realities – are constituted by such conventions (Bickhard, 2008, 2009b). These are conventions concerning the interactive nature of the situation – *situation conventions*.

[13]Social realities range from momentary common understandings among participants to a conversation, to two oncoming pedestrians passing each other on the right, to a check out clerk relationship to a customer, to institutions of government, to language, to friendships, and so on. There are many kinds of such realities, and, generally, many instances of such kinds.

[14]For some discussion of important differences in the underlying model of convention from that of Lewis, see Bickhard (2008, 2009b).

3.5.2 Conventions

Lewis developed his model of convention most centrally in order to account for language as conventional, and he also generalized the model to other kinds of what might be called *institutionalized conventions* – those that repeat across people and occasions, such as driving on the right hand side of the road. I do not disagree with the characterization of such phenomena as conventional, but the model offered here arises not from considerations of language, though it has much to say about language (see below), but rather as a kind of fixed point problem that arises intrinsically whenever complex agents encounter each other. The problem is to find a fixed point, one that is relatively coherent and thus successful in the mutual situation knowledge anticipations, in the mutual characterizations of the agent-containing situations. This involves some changes in Lewis's underlying model, changes which have some potentially important consequences.

For a first point, not all situation conventions have to be institutionalized. In every conversation, for example, at any given point (assuming that the conversation is succeeding in being mutually intelligible) there is a mutual characterization of the situation among the participants to the conversation concerning the current status of the conversation – a kind of ongoing common knowledge. This ongoing, continuously updated, common understanding is what permits, for example, common resolution of pronouns, and, thus, maintenance of the ongoing common knowledge. The apperceptions of the participants to the conversation must maintain the coherence that constitutes the situation convention, if the conversation is to not fail in some way. Such momentary situation conventions may have never occurred before, and may never again. Such highly *non*-institutionalized situation conventions cannot be modeled with Lewis's model based on *regularities of behavior* – there are no repetitions across which to be regular, and the crucial coherences are among interactive characterizations, not among behaviors.

Another difference from Lewis's model is a more central involvement of normativity. One criticism of Lewis's model is that it does not account for the normativities of conventions and of upholding them among participants (e.g., Gilbert, 1989). There is a kind of normativity in that participants have a common interest in arriving at a solution to the underlying coordination problem, and that interest constitutes a form of normativity, but the convention itself, defined in terms of regularities of behavior, either exists or it doesn't. There seems to be simply a matter of fact there, without any strong normativity. Driving on the right hand side of the road, however, seems to involve a strong normativity, not just a conventional convenience. Similarly, presenting oneself in the role of teacher or judge is normative, not just a convenient solution to a coordination problem about how to act.

In the model outlined above, situation conventions arise not just as regularities of behavior, but as solutions to the basic problem of how to understand one's situation, and one's place – and others' places – in that situation. Such characterizations, in general, are central to one's functioning in the world, and one's place in that world. This is a much stronger sense of normativity in social situations.

For example, it is essential that one's *presentation of oneself* as a competent, reliable, social interactor with sufficient integrity to support the flow of situation conventions *be accepted by others* in order to function socially at all. Again, this is a much more central involvement of normativity than can be found in Lewis's model.[15]

3.5.3 Institutionalized Conventions

Social realities in the moment are constituted by situation conventions, but social realities across society are built on *institutionalized* conventions. Driving on the right hand side of the road is one example, but there are multiple other kinds of institutionalization that are important.

For example, the anticipative interactive coherences that constitute a situation convention may involve different anticipations for differing participants in the situation. They may be understood by all involved to be going to be engaged in differing courses of further interaction: one might drive animals toward another, while that other hides and waits for the opportunity to kill an animal and share it with the group. Such interrelated kinds of anticipations can constitute roles, and typifications of such interrelations can constitute types of roles (Berger & Luckmann, 1966). Uncles, for example, might be expected to engage in certain kinds of activities and stances with respect to nieces and nephews, but the role of "uncle" might also become typified such that a person who is not a biological uncle could take on the role of Uncle.

It was and is a major development when a society develops the notion of an institutional role in terms of its relations to other roles, and distinct from the person who might at a given time occupy that role. This is the foundation for bureaucracy, which, for all of its inefficiencies and frustrations, is much more efficient and powerful than individual (e.g., feudal) or biologically (e.g., clan) based social organization.

The model of situation conventions, thus, can address social realities from momentary common understandings and anticipations to broad social institutions. That includes, so I argue, language.

[15]Another criticism of Lewis's model is that it depends on a rather rigid model of rationality, and of unrealistic assumptions about how rational thought works (Gilbert, 1989). I will not present the interactive model of rationality here, but suffice it to say that, whether or not those criticisms are valid against Lewis, this model does not involve such models of rationality (see Bickhard, 2002, 2008).

3.5.4 How Are Situation Conventions Created and Evoked?

How do conventions emerge; how are they created or modified? One important answer is: By precedence and habituation. Suppose we happen to meet at a restaurant next Tuesday and have lunch together. It is a pleasant lunch, and the following Tuesday, perhaps we both go back to that restaurant and have lunch together again. Some Tuesday hence, after meeting at that restaurant several times, we will have a convention between us – a mutually coherent organization of our anticipations – that we meet at that restaurant on Tuesdays for lunch. Crucially, we need never have talked about or negotiated this arrangement in order for it to emerge and constitute a convention.[16]

Lunch on Tuesday conventions are evoked by the arrival of Tuesdays. Many conventions, however, are evoked by activities on the part of one or more participants, which may or may not have calendar or other kinds of contextual constraints. I may, for example, invoke the convention of walking to the right of oncoming pedestrians by tending slightly to the right. That is, I do something that *presupposes*, and thereby evokes, the convention to be invoked. If this is understood and accepted by the oncoming people, we act in accordance with that convention. The *presupposition-in-action* yields the convention presupposed.

Of course, I may be in a country in which it is customary to walk to the left, or the oncoming person may misunderstand my tending to the right "gesture" as indicating, for example, that I am heading into a store. The invocation of conventions is by presupposition of those conventions, but the epistemology about such conventions is inherently tentative and fallible.

Note that the sense in which the mutual anticipations are coherent is an implicit sense. It need not involve any explicit representations of the fact that any such convention exists, or even that conventions as a "kind" exist. What is required is a coordinative framework of anticipations that mutually works as those anticipations are engaged. What prevents or annihilates convention is a failure of the anticipations to be interactively coherent, or, in some more sophisticated cases, an explicit negation on the part of one (or more) of the parties involved about the anticipations to be played out – such as a con man who wishes to deliberately invoke certain social realities that he knows are not correct (do not actually exist) because there is an explicit intention on his part, thus anticipation, to violate that anticipatory framework at some opportune time.

Convention, in these senses, does not require explicit representation of convention or of the situation as being conventional. It is because of this sufficiency of implicitness that even infants (and adults) can participate in conventional interactions, such as peek-a-boo, without having to think through any layers of "I know

[16]This kind of possibility for the establishment of convention is of central importance for Lewis (1969), who wants to characterize language as conventional: if the only way to establish a convention is via negotiation, then what is the language in which the conventions that constitute language are negotiated?

that you know that I know …”. What is required is interactive characterization coordinative coherence *in fact*, and a lack of doubt or negation (at any reflective level) of such coherence.

Convention emerges, when it does, in processes of apperception concerning social situations. Apperception is the process of updating and maintaining “knowledge” of the interactive nature of a situation: apperception can and does change prior or ongoing situation knowledge.

So, convention emerges in apperceptive processes, processes that are evoked by, for example, actions that presuppose convention. Actions may also evoke processes that *change already existing convention*: action can evoke apperception, which updates situation knowledge, and, if social participants’ situation-knowledges are updated in a coherence maintaining fashion, some new convention will now obtain. Still further, such convention altering effects can be themselves conventionalized: all convention invocation is apperception invocation, and apperception is inherently context dependent, so when that context dependency is itself conventionalized, we have conventional means for invoking or changing convention.

Insignia of rank, for example, evoke a conventional framework of expectations and obligations in contexts in which those insignia are relevant and operative. They change the respective situation knowledges of the participants. An even more powerful tool for such alterations of and operations on situation conventions is language (see below).

3.5.4.1 Emergence and Implicitness

It is worth further emphasizing here two properties of convention in this model: emergence and implicitness. It is because of the possibility of *implicitness*, for example, that hierarchies of reflective thoughts are not required for the existence of conventions and conventionalized processes – for infants, toddlers, and adults.

It is because of the *emergent* nature of convention that no basic primitive social ground has to be posited, such as we-intentionality (Searle, 1995) or the we-mode (Tuomela, 2013). In both cases, these are not derivable from the “I”-version. These are explicitly *not* emergence models, and one important reason why they cannot be is because of the background assumption that representation and reason are necessarily explicit, based generally on explicit propositional encodings (see also Bratman, 2007; Gilbert, 1989), and there is no way to account for the emergence of foundational encodings (Bickhard, 1993, 2009b). Implicitness makes possible the emergence of *in fact* interactive anticipatory coordinative coherence without anyone having to know or explicitly represent that is what is going on, let alone to think through some hierarchy of reflections on such “knowledge”.

3.5.5 A Word About Language

The model outlined here does not and cannot support standard encodingism frameworks – usually of an information semantic form, though not necessarily – that posit information "coming in" via perception, being processed in cognition, and then yielding outputs in the forms of action and utterances. That is, it cannot support standard approaches to perception, cognition, or language (Bickhard, 2009b).[17] Here I would like to say a bit more about how this model nevertheless addresses language – arguably with much more success than standard approaches, though elaborating that claim must be reserved for elsewhere.[18]

In this model, utterances modify social realities, via apperceptions of those utterances – where social realities, as outlined above, are constituted as coordinative relations among various social participants' interactive characterizations of their situation. These characterizations, in turn, are constituted as representational anticipations of interactive possibilities: thus, utterances *modify* representations, but do not *transmit* them via encodings. They "operate on" conventions (via apperception) about the current situation.

Social realities are solutions to coordination problems concerning what future possibilities might be anticipated. In human social realities, those future potentialities are massively, though not entirely, constituted in potentialities for further language. Thus, language is ontologically a major part of those social realities.

Language, then, functions in accordance with conventions concerning the apperceptive effects of utterances and parts of utterances (Bickhard, 1980). In a broad sense, it is an institutionally conventionalized toolbox for the construction of utterances which have conventionalized effects on situation conventions.

3.5.5.1 Sub-utterances and Grammar

A "full" utterance in this model elicits a full transformation of a situation convention. Partial or sub-utterances can create (*linguistic* situation convention) circumstances in which other kinds of sub-utterances can function. For example, one partial

---

[17]Information semantics is the currently dominant framework within psychology, cognitive science, and philosophy. It constitutes a fundamental equivocation between technical information – a condition of being correlated with – and semantic, or representational, information. If X is correlated with Y, then knowing X can permit inferring something about Y – but only if X and Y and the correlation are already known and represented. If all of these are known, then X can be used as an encoding of (properties of) Y, but, as usual, encodings require that all of the relevant representations be already available. Correlational information cannot, in itself, constitute representation. This point is relatively well known, though often ignored, but what else is required has no resolution. The problem of organism-detectable error, discussed above, is one of several reasons why it *cannot* be resolved. See Bickhard (1980, 1993, 2009b, 2014) for further discussion.

[18]For more developed discussions of language, and also of how perception can be modelled within this framework, see, for example, Bickhard (2009b; Bickhard & Richie, 1983).

utterance may differentiate some part or aspect of the overall commonly understood (or at least commonly and coordinatively anticipated) situation convention, making it a focus for the second part, while the second partial utterance might modify that part or aspect. In "The ball is red", for example, "the ball" will, if successful in creating and maintaining situation convention, differentiate a ball that is represented in common among the participants, while "is red" will, perhaps, modify those representations to include the property of being red, or at least will bring into the common knowledge of the situation convention the fact or assumption that the ball is red. By differentiating *types* of incomplete utterances in this manner, a grammatical framework emerges that has important similarities to categorical grammars (though also some important differences) (Bickhard & Campbell, 1992).[19]

3.5.6 Persons

There is a special emergence, according to this model, of social persons. Persons are agents, and they are social agents, but their sociality is fundamentally different from that of, for example, social insects. Ants are social, and they are emergently social, but the sociality is at the level of the ant nest as a whole. There is no social emergence at the level of individual ants.

In contrast, persons are intrinsically social agents at an individual level. Human infants are intrinsically open to developing as social persons, but the manner in which they do so and the particular agents that they become is deeply dependent on familial, social, and cultural contexts (Berger & Luckmann, 1966). Human infants become social persons within that context in the sense that they become agents that can participate in the social process and realities of that family, society, and culture – and thereby participate in the constitution of those social realities.

Persons are developmentally emergent social agents: agents that have developed as participants in socio-cultural social realities, and that thereby co-constitute those social realities. Their ontology is at the same level as human sociality per se. In particular, they constitute an emergent kind of process within the general framework of complex agents (Bickhard, 2013).

This is important for the current discussion in that it entails that persons are intrinsically involved, ontologically involved, in the normativities of social processes and realities.

[19]Note also some similarities here to file-change models, though without any assumptions of or restrictions to explicit propositional encodings (Heim, 1983; Lewis, 1979).

3.6 Stakes in Sociality

Persons, then, are ontologically constituted as socio-cultural interactive agents. As such, persons have normative stakes in their sociality – for example, successful functioning in social realities requires sufficient establishment and maintenance of those social realities, which, in turn, requires presenting oneself as – and being accepted as – reliable, competent, and trustworthy social agents. Successful functioning in social realities constitutes the processual existence of the social person, similar to the sense in which "successful" burning constitutes the existence of a candle flame – a strong stake indeed.

3.6.1 Sociality and Historicity

Social realities require constant maintenance, repair, and development. To be a person is, among other things, to be able to participate in such maintenance, repair, and development. This requires the ability to respond to failures of social realities, ranging from misunderstandings to encounters that go beyond current social realities, such as maintaining and developing a relationship in encountering a crisis for one or more parties to that relationship – or, of importance for current purposes, in unfolding further emergent aspects and characteristics of social realities, such as friendships. That is, social realities necessarily have histories of development – and deep social realities, e.g., long standing friendships or marriages, have deep histories of development – "deep" in the sense that more and more aspects of the emergent person are dependent for their interactive realization on those friendships or partnerships or other kinds of relationships.

Such relationships are themselves conventions, generally between two or a few people, that have developed and are developing over longer time frames and with respect to more central concerns and values of the persons involved. The historistic development of such relationships requires values and normative stakes with respect to which the learning and development of the relationships can occur – to constitute criteria of appropriateness and inappropriateness, success and failure, in the further development of a relationship, and, thus, of the persons involved. The relationships are in important respects the conventions of such *development*, not just of particular interactive common understandings in the moment.

In sum, such learning requires normative stakes relative to which learning can occur, and, thus, such developmental historicities can be realized.

3.7 Robots and Sociality

Robots can simulate complex forms of social interaction, and they could be programmed to detect various pre-designated kinds of "failures" and to use various heuristics to recover and learn from them – a simulation of normativity, and of the learning and development that is based on such normativity. But they have no intrinsic stake in the world, nor in their existence in the world, nor in their existence as social agents.[20] Thus, they can only simulate interaction forms and criteria for failures that have been anticipated in their design. They are limited to what their design can anticipate. In that sense, they cannot have genuine developmental historicity, thus not genuine social being.

3.7.1 Is This Limitation Necessary?

Could, however, artificial agents be developed that *do* have intrinsic sociality and its normativities? I would argue "Yes", but they would have to have inherent emergent stake in their existence as social agents, which, in turn, requires having a stake in their existence in the world.

To accomplish this would require, among other things, a model of emergent normativity as it occurs in biology, and human beings in particular. An artificial agent, then, could presumably realize or instantiate such normativity.

As discussed above (and, e.g., Bickhard, 2009b), this requires certain kinds of far-from-thermodynamic-equilibrium processes. That is, normativity emerges most primitively as normative function, and normative function emerges in certain kinds of processes that are far from thermodynamic equilibrium. Far-from-equilibrium processes have a stake in the world in the sense that their existence requires maintenance of the far-from-equilibrium conditions that existence depends on. Their existence is at stake in whether or not those far from equilibrium conditions are in fact maintained. Robots, as we currently understand them, do not realize such processes – they are (primarily) not far from equilibrium – but artificial systems could, in principle, realize such far-from-equilibrium processes, and, thus, realize emergent normativities of various sorts.[21]

[20]With no stake in their existence in the world, they cannot have a stake in their existence in the world as social agents.

[21]For further discussion of this issue regarding artificial agents, see Bickhard (2009a).

3.7.2 Robots as Moral Agents

The issue of having a stake in their own existence is also of relevance for moral issues. In particular, this would seem to be at the center of being an object of moral regard – a moral patient. Moral regard is, in this sense, a kind of respect for the stake that other beings have in their own existences. Having such a stake, however, is not, in general, sufficient for being a moral *agent*. For example, an infant has a stake in the world, and in its own existence in the world, and, thus is an object of moral regard, but it is not (yet) a moral agent. A robot, thus, that has a stake in its own existence would be a an object of moral regard; there would be, perhaps, moral responsibilities toward such beings. But they would not just in virtue of that be moral agents, agents with moral responsibility. What is further required for moral agenthood must be addressed elsewhere.

3.8 Conclusion

There are multiple reasons for working with a process metaphysics, and multiple important consequences to doing so. Among these consequences are the possibilities of emergence, and, in particular, normative emergence. With an initial emergence of normative function in far-from-thermodynamic-equilibrium processes, a directly related model of the emergence of representation, with its normative dimension of truth and falsity, is easily constructed. This is in terms of the normative characteristics of an agent's indications or anticipations of what it could do next, and the possibility that those indications or anticipations might be false – indicating that some interactive possibility holds presupposes that sufficient supporting conditions hold, and they might not.

With such an inter-action based model of representation and cognition, a special problem arises when two (or more) complex agents are in each other's presence: characterizing the situation requires characterizing the other agent, including that agent's characterization of the first agent, which includes the first agent's characterization of the second, and so on. Insofar as both agents have an interest in arriving at a successful characterization, this constitutes a coordination problem in the sense of Schelling, and, thus, any solution to this problem constitutes a convention in a (broadened) sense of Lewis. I argue that conventions, modeled in this broader manner, capture all social realities, from the momentary to the institutional to language to the deep, such as a friendship. One important characteristic is that the persons who participate in and constitute such social realities have normative stakes in their participations, and in their existence and acceptance as agents who are capable of the creation, maintenance, and development of such conventions. This is especially important for social processes, such as friendships, that develop over longer time periods and engage central values of the persons involved.

It is such stakes that current robotics does not capture, and, thus, contemporary robots cannot be genuine social agents. Robots as currently designed are limited to various kinds and degrees of simulation of sociality. For some purposes, that may suffice. They cannot be genuine social and moral agents, however, without having an intrinsic stake in their existence as social agents. Among other consequences: they cannot, in contemporary form, be genuine friends. That will require different kinds of design and development principles than are currently available.

Acknowledgements This is an expanded version of Bickhard, M. H. (2014). Robot Sociality: Genuine or Simulation? In J. Seibt, R. Hakli, M. Nørskov (Eds.) *Sociable Robots and the Future of Social Relations*. Series: Frontiers in Artificial Intelligence and Applications. (81–86). Amsterdam: IOS Press.

References

Aitchison, I. J. R. (1985). Nothing's plenty: The vacuum in modern quantum field theory. *Contemporary Physics, 26*(4), 333–391.

Aitchison, I. J. R., & Hey, A. J. G. (1989). *Gauge theories in particle physics*. Bristol/England: Adam Hilger.

Allen, J. W. P., & Bickhard, M. H. (2013). Stepping off the pendulum: Why only an action-based approach can transcend the nativist-empiricist debate. *Cognitive Development, 28*, 96–133.

Berger, P. L., & Luckmann, T. (1966). *The social construction of reality*. New York: Doubleday.

Bickhard, M. H. (1980). *Cognition, convention, and communication*. New York: Praeger Publishers.

Bickhard, M. H. (1993). Representational content in humans and machines. *Journal of Experimental and Theoretical Artificial Intelligence, 5*, 285–333.

Bickhard, M. H. (2002). Critical principles: On the negative side of rationality. *New Ideas in Psychology, 20*, 1–34.

Bickhard, M. H. (2003). Variations in variation and selection: The ubiquity of the variation-and-selective retention ratchet in emergent organizational complexity, part II: Quantum field theory. *Foundations of Science, 8*(3), 283–293.

Bickhard, M. H. (2005). Consciousness and reflective consciousness. *Philosophical Psychology, 18*(2), 205–218.

Bickhard, M. H. (2008). Social ontology as convention. *Topoi, 27*, 139–149.

Bickhard,M. H. (2009a). The biological foundations of cognitive science. *New Ideas in Psychology, 27*, 75–84.

Bickhard, M. H. (2009b). The interactivist model. *Synthese, 166*, 547–591.

Bickhard, M.H. (2013). The emergent ontology of persons. In J. Martin & M. H. Bickhard (Eds.), *The psychology of personhood: Philosophical, historical, social-developmental, and narrative perspectives* (pp. 165–180). Cambridge: Cambridge University Press.

Bickhard, M. H. (2014). What could cognition be, if not computation ... or connectionism, or dynamic systems? *Journal of Theoretical and Philosophical Psychology, 35*(1), 53–66.

Bickhard, M. H. (in preparation). *The whole person: Toward a naturalism of persons—Contributions to an ontological psychology*.

Bickhard, M. H., & Campbell, R. L. (1992). Some foundational questions concerning language studies: With a focus on categorial grammars and model theoretic possible worlds semantics (target article for commentary in a special double issue). *Journal of Pragmatics, 17*(5/6), 401–433.

Bickhard, M. H., & Richie, D. M. (1983). *On the nature of representation: A case study of James Gibson's theory of perception*. New York: Praeger Publishers.
Bratman, M. E. (2007). *Structures of agency*. Oxford: Oxford University Press.
Brown, H. R., & Harre', R. (1988). *Philosophical foundations of quantum field theory*. Oxford: Oxford University Press.
Campbell, D. T (1974). Evolutionary epistemology. In P. A. Schilpp (Ed.), *The philosophy of Karl Popper* (pp. 413–463). La Salle, IL: Open Court.
Campbell, R. J. (1992). *Truth and historicity*. Oxford: Oxford University Press.
Campbell, R. J.(2011). *The concept of truth*. New York: Palgrave Macmillan.
Campbell, R. J., & Bickhard, M. H. (2011). Physicalism, emergence and downward causation. *Axiomathes, 21*, 33–56.
Cao, T. Y (1999). *Conceptual foundations of quantum field theory*. Cambridge: University of Cambridge Press.
Chang, C. C., & Keisler, H. J. (1990). *Model theory*. Amsterdam/New York: North Holland.
Christensen, W. D., & Bickhard, M. H. (2002). The process dynamics of normative function. *Monist, 85*(1), 3–28.
Davies, P. C. W.(1984). Particles do not exist. In S. M. Christensen (Ed.), *Quantum theory of gravity* (pp. 66–77). Bristol, England: Adam Hilger.
Dewey, J. (1960). *The quest for certainty*. New York: Capricorn Books. (Originally published in 1929).
Fodor, J. A. (1981). The present status of the innateness controversy. In *Representations: Philosophical essays on the foundations of cognitive science* (pp. 257–316). Cambridge, MA: MIT Press.
Fodor, J. A. (1990). Information and representation. In P. P. Hanson (Ed.), *Information, language and cognition* (pp. 175–190). Vancouver: University of British Columbia Press.
Gilbert, M. (1989). *On social facts*. Princeton, NJ: Princeton University Press.
Gill, M. L. (1989). *Aristotle on substance*. Princeton, NJ: Princeton University Press.
Graham, D. W. (2006). *Explaining the cosmos*. Princeton, NJ: Princeton University Press.
Guthrie, W. K. C. (1965). *A history of Greek philosophy* (The presocratic tradition from Parmenides to Democritus, Vol. II). Cambridge: Cambridge University Press.
Hale, B., & Wright, C. (2000). Implicit definition and the a priori. In P. Boghossian & C. Peacocke (Eds.), *New essays on the a priori* (pp. 286–319). Oxford: Oxford University Press.
Halvorson, H., & Clifton, R. (2002). No place for particles in relativistic quantum theories? *Philosophy of Science, 69*(1), 1–28.
Heim, I. (1983). File change semantics and the familiarity theory of definiteness. In C. S. R. Bauerle & A. Stechow (Eds.), *Meaning, use, and interpretation of language* (pp. 164–189). Berlin: de Gruyter.
Huggett, N. (2000). Philosophical foundations of quantum field theory. *The British Journal for the Philosophy of Science, 51*(supplement), 617–637.
Hume, D. (1978). *A treatise of human nature*. Oxford: Oxford University Press. (Index by L. A. Selby-Bigge; Notes by P. H. Nidditch).
Kuhlmann, M., Lyre, H., & Wayne, A. (2002). *Ontological aspects of quantum field theory*. River Edge, NJ: World Scientific.
Lewis, D. (1969). *Convention*. Cambridge, MA: Harvard University Press.
Lewis, D. (1979). Scorekeeping in a language game. *Journal of Philosophical Logic, 8*, 339–359.
Millikan, R. G. (1984). *Language, thought, and other biological categories*. Cambridge, MA: MIT Press.
Millikan, R. G. (1993). *White queen psychology and other essays for Alice*. Cambridge, MA: MIT Press.
Piaget, J. (1954). *The construction of reality in the child*. New York: Basic.
Piattelli-Palmarini, M. (1980). *Language and learning: The debate between Jean Piaget and Noam Chomsky*. Cambridge, MA: Harvard University Press.
Schelling, T. C. (1963). *The strategy of conflict*. New York: Oxford University Press.

Schurz, G. (1997). *The is-ought problem: An investigation in philosophical logic* (Trends in logic, Vol. 1). Dordrecht/Boston: Kluwer Academic.
Sciama, D. W. (1991). The physical significance of the vacuum state of a quantum field. In H. R. B. S. Saunders (Ed.), *The philosophy of vacuum* (pp. 137–158). Oxford: Clarendon.
Searle, J. R. (1995). *The construction of social reality*. New York: The Free Press.
Tiles, J. E. (1990). *Dewey*. London/New York: Routledge.
Tuomela, R. (2013). *Social ontology: Collective intentionality and group agents*. Oxford: Oxford University Press.
Weinberg, S. (1977). The search for unity, notes for a history of quantum field theory. *Daedalus, 106*(4), 17–35.
Weinberg, S. (1995). *The quantum theory of fields* (Foundations, Vol. 1). Cambridge: Cambridge University Press.

Chapter 4
Prioritizing Otherness: The Line Between Vacuous Individuality and Hollow Collectivism

David Eck and Alex Levine

Abstract In conceptualizing social robotics it is tempting to view the social dimension as simply an additional layer, overlaid upon the motility and autonomy of individual robots. This paper problematizes such an approach, arguing instead for treating the process of *interaction* as, in a sense, prior to both individuality and sociality. After analyzing the notion of priority at work in claims of the form, "individuality is prior to sociality," we turn to lacunae in enactivist approaches to sociality exemplified by the work of De Jaegher and Di Paolo. These, we argue, can only be overcome by means of an interdisciplinary appreciation of the *problem of alterity*.

Keywords Ontological priority • Epistemic priority • Enactivism • Social robotics • Participatory sense-making • Alterity • Embodiment

4.1 Introduction

Motility is one of the defining features of robots, and autonomy one of the primary goals of robot design. Motility and autonomy are also often touted as fundamental characteristics of many clades of organisms, and especially of animals. It is thus not surprising that robotics has always been informed by disciplines devoted to the study of animal motility and autonomy. Of course there are well known reasons for treating zoological analogies with caution, among them the fact that motility and autonomy, among other relevant features of animals, remain in many ways poorly understood. In the emerging field of social robotics, such caution should be redoubled. If a social robot is one that interacts with humans and other robots, it is tempting to think of such interactions as simply a new layer of engineering

D. Eck (✉)
Humanities & Social Sciences Division, Cañada College, 4200 Farm Hill Blvd, 94061, Redwood City, CA, USA
e-mail: eckd@smccd.edu

A. Levine
University of South Florida, 4202 E. Fowler Ave, FAO 226, 33620, Tampa, FL, USA
e-mail: levineat@usf.edu

R. Hakli, J. Seibt (eds.), *Sociality and Normativity for Robots*, Studies in the Philosophy of Sociality 9, DOI 10.1007/978-3-319-53133-5_4

and dynamics "added on" to autonomous robots. In this paper, we argue that such an approach to robot sociality, whether adopted explicitly or simply presupposed, is likely to prove misguided, on the grounds that the analogous approach to human sociality is demonstrably misleading. The claim that individuality is prior to sociality—and the less common converse assertion, that sociality is prior to individuality, founder on fundamental conceptual grounds, beginning with an ambiguity in the notion of *priority*. But one must start somewhere. In the tradition of process metaphysics, we argue that the place to start is not the individual, nor the collective, but rather the process of *interaction*.

In his classic paper "Intelligence without Representation," Rodney Brooks points to the persistent failure of AI researchers armed with the most sophisticated symbolic and connectionist models of representation to build intelligent systems. He surmises that perhaps, "Representation is the wrong unit of abstraction in building the bulkiest parts of intelligent systems" (Brooks, 1991, p. 139). Here, by analogy, we want to suggest that perhaps the atomic human subject is the wrong sort of entity to presuppose in building social systems. Treating it as such can lead to the vacuous individuality of atoms that, while numerically distinct, might in most respects be interchangeable. But it is not enough to simply turn the tables, and to begin with the social instead. To do so is to run the risk of exchanging vacuous individuality for hollow collectivism.

The philosophical tradition offers us numerous variations on the claim sometimes made explicitly, more often implicitly presupposed—that for human agents, individuality is prior to sociality. As an implicit presupposition, it vitiates attempts to account for sociality, but also undermines the category of the individual person itself. We will call it the priority of individuality thesis, or POI. In the first half of this paper, we wish to problematize that thesis. In the second, we turn to the promise of an enactivist framework for understanding and examining individuals and social systems. The challenges it faces are the same as those facing POI, but turned inside out. Nonetheless, we believe that recent extensions to the enactivist notion of sense-making have the potential, with sufficient care, to avoid both sorts of challenge. We conclude by making this case.

We begin by challenging the claim that individuality must be understood as prior to sociality on two grounds. First, in most of its variants, POI turns on an equivocal notion of *priority*. There are at least three distinct kinds of priority at work in the philosophical literature: epistemic priority, ontological priority, and diachronic temporal priority. Equivocations between either or both of the first two (epistemic and ontological) and the third are further vitiated by the fact that diachronic temporal priority is itself not unequivocal. *A* can be temporally prior to *B* on any one or several of a range of overlapping timescales.

Second, we consider two examples from the philosophical canon in which POI can be seen to have pathological, even paradoxical consequences. By individuating moral agency by recourse to *autonomous individual human subjects*, both Kantian and utilitarian accounts of morality erase the possibility of morally significant differences among such subjects. In consequence, paradoxically, the autonomous individual human subject evaporates as a basis for individuating moral agents.

As a further example, we consider two classic solutions to the problem of other minds, what Churchland (2013, Chap. 4) calls the argument from analogy and the hypothetico-deductive argument. Both presuppose that the problem of other minds is the epistemological problem of identifying individual subjects as such. As Merleau-Ponty has suggested (2012, pp. lxxv–lxxvi), both "solve" the problem of other minds by essentially erasing otherness; there may be multiple instances of mind, but *qua* instances of mind, there are no significant differences between them.

Having articulated and defended our critical thesis, we sketch an alternative approach to examining social interaction by drawing upon enactivism both as an explanatory framework and as a case study of the complications in overcoming the POI thesis. With regard to the former aim, we explicate De Jaegher and Di Paolo's concept of participatory sense-making (2007) and note some of its basic insights on social interaction. We then critique De Jaegher's attempts (2013) to scale participatory sense-making to larger social contexts: her ascription of positive and negative meanings to various patterns of social interaction entails a vacuous collectivism, a mirror image of the very atomistic individualism that participatory sense-making undermines. Having critiqued De Jaegher's expansion of participatory sense-making, we briefly consider an alternative way of extending the concept. While perhaps more modest in its explanatory ambitions, our approach to understanding how participatory sense-making shades into larger contexts captures the priority of interaction without thereby annihilating the role of the particular participants. Hence, while our critique of the POI thesis is directed primarily against advocates of atomic individualism, it counsels against any social theory that dilutes or erases the otherness intrinsic to interaction.

4.2 Getting Priorities Straight

The philosophical canon is replete with claims of the form, "A is prior to B." This is not the place for a systematic survey, but for present purposes, a single example will suffice. Consider the contentious, but not uncommon claim, that language is prior to thought. Some interpreters of Hegel's Aesthetics have read this assertion into his thesis that "poetry precedes philosophy." It has also been attributed to Ferdinand de Saussure. The thesis is explicitly expressed in Wilfrid Sellars's (1956) essay, "Empiricism and the Philosophy of Mind." Sellars's version of the thesis is qualified; language precedes thought in the Aristotelian or Thomistic "order of knowing" (Sellars, 1956). Jerry Fodor's argument in *The Language of Thought* (Fodor, 1975) turns on the premise that there is "No computation without representation." On this view, representation, which is structured in many ways like a language, is prior to thought, a form of computation. Michael Dummett claims that "an account of language yields an account of thought, and there is no other means by which an account of thought may be given" (Dummett, 1981, p. 39). Donald Davidson, while insisting that "neither language nor thinking can be fully explained in terms of the other, and neither has conceptual priority," argues that "a creature cannot have thoughts unless it is an interpreter of the speech of another" (Davidson, 1984, p. 156). Finally, Robert Brandom may be adopting a version of this thesis in his 1994 book, *Making it Explicit*. Please note that we are not asserting that all of these figures are making the same claim. Nor do we think that, insofar as it is fair to say that each advocates a version of the claim that "language is prior to thought," they all fall prey to every one of the possible equivocations outlined below. Some, including Sellars and Davidson in the passages we have cited, take pains to qualify their theses in ways that clearly preclude at least *some* of these equivocations. But we believe, though we cannot argue it here, that they all commit at least *one* of them.

Disentangling different senses of priority, we begin with *epistemic* priority. We will say that *A* is epistemically prior to *B* when, for a given epistemic subject in a given situation, knowledge of *A* is a necessary precondition for knowledge of *B*. Note that it is essential to relativize epistemic priority to a given subject in a given situation. We take very seriously Philip Kitcher's analysis of mathematical apriority, developed as part of his case for mathematical empiricism in *The Nature of Mathematical Knowledge* (Kitcher, 1984). Kitcher argues, persuasively in our view, that the distinction between apriority and aposteriority designates a difference in degree, not a difference in kind. What is a priori for one subject in a given situation may be a posteriori for another subject in another situation, and vice versa. In the absence of any persuasive case for the existence of a class of judgments that are *a priori tout court*, regardless of epistemic subject and situation, epistemic priority always implies an element of diachronic temporal priority; *A*'s being epistemically prior to *B* requires that *A* be known *before B*. This is the priority of Aristotle's, Thomas Aquinas's, and Wilfrid Sellars's "order of knowing" (Sellars, 1960).

This said, it is tempting to identify *ontological priority* with the Thomistic "order of being", but to the extent to which the order of being involves an element of *causal priority*, it is instead captured under what we shall shortly be calling *diachronic temporal priority*. It would also be misleading to say only that *A* is *ontologically prior* to *B* when the existence of *B* requires (or entails) the existence of *A*. For the existence of a triangle requires or entails the existence of three angles summing to 180°, but there seems to be no compelling reason to take the three angles as ontologically prior to the triangle. Ontological priority, like other relations of priority, must be asymmetrical. So we will say instead that A is ontologically prior to B when the existence of B entails the existence of A, but the existence of A does not entail the existence of B. This may be roughly the "conceptual" priority Davidson disavows when he says that neither language nor thought "has conceptual priority," for the right sort of asymmetry is absent (Davidson, 1984, p. 156).

The *diachronic temporal priority* of A over B is given when A precedes B in temporal sequence, which sounds simple enough. But there are many claims of the form "A is prior to B" for which the duration of the interval of temporal priority separating A from B matters rather a lot. Any claim of the form, "A causes B" implies the diachronic temporal priority of A over B, but on almost any account, causality comes in various shades of proximal and distal. The claim "hydrogen is prior to the heavy elements" suggests priority on a cosmological timescale. "Metazoa are prior to primates" invokes priority at the long end of the evolutionary timescale, while "agriculture is prior to adult human lactose tolerance" targets the intersection of evolutionary and historical timescales. "Dry rice cultivation was prior to wet rice cultivation," "Latin was prior to Italian," and "The iPhone 5 was prior to the iPhone 6," are all examples of priority claims that make sense on various historical timescales. Then there is the ontogenetic timescale, the timescale on which an individual organism develops. For example, "the blastocyst is prior to somatic differentiation," "deciduous teeth are prior to permanent teeth," or "adrenarche is prior to menarche."

This is by no means intended as an exhaustive survey of all the different senses of priority still current in the philosophical literature. But as a glance at the ambiguities inherent in our examples will show, it is enough to go on. For Fodor, the claim that language is prior to thought, or its equivalent, has both temporal and ontological elements. As a self-confessed "mad-dog" nativist, he holds that the language of thought—the internal representational system in which cognitive computation takes place—must be present at birth, or else learning by hypothesis formation and confirmation could never take place. But the slogan, "no computation without representation" is conceptual or ontological, not temporal. Dummett's claim, "an account of language yields an account of thought, and there is no other means by which an account of thought may be given," appears at first blush to be asserting the epistemic priority of language over thought, for it is the *account* of language that is prior to the *account* of thought (Dummett, 1981, p. 39). Its giving rise to the account of thought presumably takes place in time, though it is not clear from Dummett's text which timescale or timescales apply—perhaps the timescale of the individual learning trajectory, the institutional-historical timescale of the philosophical enterprise, or some combination of the two. Finally, for Dummett and those of similar philosophical proclivities, the conditions necessary for giving an account of something are constitutive of the thing itself. Seen in this light, the priority of language over thought has an ontological or conceptual dimension here, too.

In short, as the claim "language is prior to thought" illustrates, priority claims are replete with ambiguities, and with the potential for equivocation they entail. In some instances there are arguments for the priority of one sort of priority over another, as it were. We daresay the principle of verification could be used to argue the priority of epistemic over ontological priority. But on the rare occasions when such arguments are made explicitly, they never, to our knowledge, succeed in disambiguating the notion of priority completely.

Let us now return to the supposed priority of individuality over sociality. Reading the POI thesis as asserting an epistemic priority would mean understanding it as claiming that knowledge of individuals is a necessary condition for knowledge of social systems. True or false, the epistemic POI thesis begs a number of important questions regarding the viability of the social sciences as *autonomous special sciences*. Does knowledge in sociology, anthropology, political science and the like depend on the prior completion of the science of psychology, or even on its progress? We don't know, though we suspect not. But even those more inclined than we to this kind of reductionism would do well to be wary of precluding from the outset lines of enquiry that *begin* with the social. The history of science gives us numerous examples of statistical sciences epistemically prior to the study of individual population members. Statistical mechanics illuminated the behavior of volumes of gas, understood as populations of molecules, without committing itself to claims about the properties of individual molecules. To cite an example from evolutionary biology, population genetics has done much to inform our understanding of the fitness of one or another particular individual phenotype.

Related considerations would tend to caution against an ontological POI thesis. The analogy between the body of an individual multicellular organism and a species led David Hull to challenge what he saw as the Aristotelian legacy in the philosophy of biology, the prevailing view that a species was a "spatiotemporally unrestricted class" of organisms sharing the same characteristics, where each organism was understood as (roughly) an individual substance (see e.g. Hull, 1987; see Levine, 2001 for an overview). Instead, Hull argued, species should themselves be understood as individuals, with member organisms not instances of them, but parts. Just as belonging to a particular human body is constitutive of the identity of my hand, so belonging to a particular species would be constitutive of the identity of an organism. Relative to the received ontological priority of organism over species, Hull's account inverts the constitutive dependencies. The same plausibly holds for the identity of an individual member of a eusocial insect colony. With regard to the ontological relationship between the individual human subject and the web of social relations in which he or she is embedded, there are many reasons for taking the latter as constitutive of the former. Personal identity and the choices it entails are inextricably bound up with social roles (language, socioeconomic status, job description, parental status, etc.) and with macroscale features of society and its institutions (political structures, economic relations, etc.).

As we have already suggested, many claims of epistemic or ontological priority involve an element of diachronicity, which serves as an additional source of ambiguity and complication. Let us now confine ourselves to diachronic temporal readings of the POI thesis. An idealized diachronic temporal POI thesis functions in much contractarian political theory—think of the Hobbesian individual in a state of nature, bonding together with other such for purposes of mutual protection. But even within that tradition, there is disagreement. For Rousseau, the human agent begins as a member of a family. In either case, the relevant timescale is either that of an individual human *lifespan*, or the *historical* timescale over which societies hypothetically coalesce out of the cloud of individual human constituents. If a diachronic temporal POI thesis articulated in these terms seems suspect at first blush, it only becomes more implausible the closer we look. On the *ontogenetic* timescale, in addition to itself being the product of sexual reproduction, the human infant is, of course, completely dependent for its survival on the care of older humans, typically a whole community of them. *Historical* developments also leave their mark on human ontogeny. The fact that each of us was born and raised in a society replete with written inscriptions, a society enjoying universal literacy, a phenomenon unknown before the twentieth century, has had a measurable impact on our perceptual faculties, with its counterpart in neurophysiology. According to the "cooking hypothesis" (Wrangham, 2009), the emergence in earlier hominins of the social structures necessary for the production and maintenance of fire, the storage and transportation of food, and the transmission of all of these skills, fundamentally changed the selection parameters under which human evolution subsequently took place. Each of us is, even at the level of gross anatomy, the *evolutionary* product of such social structures.

The thesis that individuality is prior to sociality is deeply ambiguous. When disambiguated, its constituent theses appear far from obvious, and in some cases, plainly false.

4.3 The Paradox of Individual Subjectivity

In this section, we approach a pathological consequence of POI, what we call the paradox of individual subjectivity, from two different perspectives: that of late modern moral theory, and that of approaches to the problem of other minds in contemporary analytic philosophy of mind. In late modern moral theory, Kantian and utilitarian approaches to moral decision-making have something important in common. Both require that the individual subject, the sole unit of moral agency for both, make decisions in a way that abstracts from interests particular to that subject. In consequence, for an individual to act morally is for him or her to elide, bracket, or erase all of the peculiarities that distinguish this individual from all others. Both Kantian moral theory and utilitarianism have shouldered a variety of burdens since their inception. But insofar as either Kantian or utilitarian theory is taken as prescribing individual actions, the consequences of both show that taking individuality as prior can make it vacuous.

On the Kantian view, to act morally is to act on maxims consistent with the Categorical Imperative. To take the second formulation of the Categorical Imperative from the *Groundwork* (Kant, 2012), this means acting only on that maxim that one could will to become a universal law of human nature. In other words, to act morally means to will that the maxim of one's action be equally binding on all individual human agents, regardless of any differences in individual perspective or interests. Each and every moral agent must (to turn to the third formulation of the Categorical Imperative) imagine him- or herself as lawgiver in the Kingdom of Ends. This view does not rule out individual difference as an empirical possibility, it simply relegates individual difference to the *merely* empirical. Morality is not, for Kant, an empirical discipline. Empirical differences among human agents are thus of no moral significance.

On the Millian rule-utilitarian view (Mill, 2002), to act morally is to act in accordance with rules that foster the greatest good of the greatest number. On this account, my own good or the good of my friends and family is, or ought to be, of no greater importance to my moral decision-making process than the good of any other. Sophisticated Millian utilitarianism does leave room for preliminary enquiries into the kinds and quantities of pleasures ideal for human beings, and may even leave room for individual difference by allowing the possibility that some pleasures may count for more in some subjects than in others. But there remains a common currency of utility, and there is no playing favorites in utility maximization. In moral terms, the utility of any individual is equivalent to that of any other.

For all the differences between them, Kantian and utilitarian approaches to individual agency construct an individual moral agent who, in order to be moral, must disavow any particularities constitutive of his or her own individuality and the individuality of anyone else. Differences do not matter because, deep down, we are all the same—man or woman, Christian, Muslim, or atheist. It is no surprise that recovering a basis for valuing diversity from this kind of universalistic account of moral agency is a persistent problem for contemporary political theory, confronted as it is with the challenges of reconciling the eighteenth century values of the Enlightenment with the twenty-first century realities of globalization and multiculturalism.

Turning to contemporary analytic philosophy of mind, we recall that its most persistent epistemological problem is the problem of other minds—the problem of knowing when one is in the presence of another mind than one's own. For ease of exposition, we will base our discussion on a textbook presentation of the potential solutions to the problem, that given in Chap. 4 of Paul Churchland's classic, *Matter and Consciousness* (Churchland, 2013). In Churchland's exposition, the two most promising approaches to the problem of other minds are the argument from analogy and the hypothetico-deductive argument.

On the argument from analogy, the individual subject begins with introspective knowledge of her own mind, assumed to be unproblematic. She then looks about her world and discovers it is full of beings who look, talk, and otherwise behave in ways similar to herself. She infers that they, too, have minds: repositories of beliefs and desires systematically related to their behavior. The problem with this approach, as Churchland points out, is that it is essentially an induction from one case—that of one's own mind—and thus very nearly the weakest sort of empirical inference imaginable.

On the hypothetico-deductive argument, which Sellars (1956) first proposed in approximately the form in which Churchland canvasses it, the individual subject instead begins by surveying her environment, taking note of the fact that it contains a multitude of beings whose behavioral complexities beggar the imagination. It is incumbent on the individual to come up with a theory that will help her cope with these beings and their behavioral complexities, a generalization suited to explaining, predicting, and controlling their behavior. Toward that end the individual hypothesizes that these beings have minds, repositories of beliefs and desires systematically related to their behavior. This hypothesis is then confirmed, or falsified and then refined, by comparing predicted behavior with observation. One problem with this approach to the problem of other minds is that it counterintuitively fails to ascribe any epistemic role to introspection. Another, related problem is that on the hypothetico-deductive approach, the individual subject is just one of the crowd; she attributes a mind to herself, as an explanatory hypothesis, in the same way she attributes minds to others.

Without discussing either in any detail, we note that, significantly, both approaches to the problem of other minds dispense with the problem rather than solving it. They dispense with it by asserting that there is nothing, or nothing interesting, that is *other* about other minds. As Jerry Fodor famously quipped, they both entail that "the psychology of individual differences is largely vacuous for our species, as it patently is for every other" (Fodor, 1981). On the argument from analogy, the individual imposes what he or she knows about herself on others; on the hypothetico-deductive approach, the individual devises a general explanatory hypothesis. Either approach allows for individuals to differ in the contents of their beliefs and desires, but nothing more.

The Kantian and utilitarian approaches to agency, and both of the aforementioned approaches to the problem of other minds, set out from an assumption of the priority of the atomic individual human subject, either as the primitive unit of moral decision-making, or as the primitive unit of cognitive subjectivity. All of the approaches we have canvassed in this section are problematic when it comes to explaining how atomic individuals participate in social relations or form social structures, because for all of them, sociality is simply an afterthought, or a bonus feature tacked on to individual subjectivity. The situation may be even worse, however, for by setting out from the priority of atomic individuality, they elide or erase individual differences as constitutive of individuality. If the POI thesis cannot account for sociality, then paradoxically, neither can it account for individuality. POI succumbs to the problems of what Maurice Merleau-Ponty, in the preface to *Phenomenology of Perception*, calls "reflective analysis."

> Reflective analysis is unaware of the problem of others, just as it is unaware of the problem of the world, because from the first flicker of consciousness it grants me the power to go toward a truth that is universal by right, and since the other is himself without *haecceity*, without place and without a body, the Alter and the Ego are one and the same in the true world, which is the unifier of minds. There is no difficulty in understanding how "I" can think the Other, because the "I" and consequently the Other are not trapped in the fabric of phenomena and have a value rather than an existence. Nothing is hidden behind these faces or these gestures, and there are no landscapes that remain inaccessible to me; there is but a shadow that owes its existence to the light. For Husserl, however, we know that there is indeed a problem of others, and the *alter ego* is a paradox. If another person is truly for-himself, beyond his being-for-me, and if we are for-each-other and not separately for-God, then we must appear to each other, we both must have an exterior, and there must be, besides the perspective of the For-Oneself (my view upon myself and the other's view upon himself), also a perspective of the For-Others (my view upon others and the view of others upon me). Of course, these two perspectives cannot be in each of us merely juxtaposed, *for then others would not see me and I would not see others*. I must be my exterior, and the other's body must be the other person himself. (Merleau-Ponty, 2012, pp. lxxv–lxxvi, original emphasis)

In the language of this text, "reflective analysis" might refer to philosophy as practiced by Cartesians, Kantians, or even the early Husserl; it certainly includes that aspect of Kantian philosophy considered in this section, a moral theory on which the individual subject has "a value rather than an existence." Furthermore, on either approach to the problem of other minds discussed above, the "Alter and the Ego are one and the same." But "for Husserl," or more precisely, for late Husserl as Merleau-

Ponty understands him, "there is indeed a problem of others." This is not a defect of the phenomenological approach he champions, but rather a virtue: other people, in all their diversity, will remain a challenge, as will the demand they pose that they be recognized for what they are, in all their alterity. The problem of other minds is a *real* problem, one confronted daily. A solution that dispenses with the problem is no solution at all.

4.4 Sociality Without Prior Individuality

We do not pretend to have made a case against any possible version of POI. But we do think we have raised sufficient doubt to motivate us to look for an account of sociality that does not take individuality as prior, in any of the myriad senses of priority we have considered. The idea of the autonomous individual subject or agent retains a great deal of the concept of an individual substance established by Aristotle. In this sense, it is one more example of the many respects in which we have yet to overcome the limitations of substance metaphysics. Treating the individual human subject as instead constituted by processes—evolutionary processes, historical processes, metabolic processes, and yes, social processes—thus recommends itself as an alternative. Constructing such an alternative will require, first, that we devise an alternative to the ontology of individual substance by articulating a sense of "process" on which processes may be understood as independent of any substantial substrate. Second, it will demand an account of both individual human subjects and the social relations in which they participate as constituted by such processes, along with a nuanced investigation of the co-constitution or co-construction of human subjects and their social relations, one that relinquishes the supposed priority of the former over the latter. Important work has been done on both of these pieces, and on putting them together. In this section, after canvassing some of this work and noting its general contours, we will turn to one specific thread of this broad and multifaceted movement in order to better assess the advances already made and some of the lingering dangers.

With regard to fundamental process ontology, a number of recent proposals suggest fruitful avenues for exploration. If the problem may be ascribed to the uncritical retention of key elements of the *Organon*, to the continued pull of individual substances, the "things" Aristotle "discovered" in his attempt to deal with problems posed by the Presocratics (Mann, 2000), we must look to other categories. One such is Seibt's category of *general process* (see e.g. Seibt, 2000, 2003), which allows "things, activities, stuffs, events, states, properties, relations, capacities, and persons, [to be] analyzed in terms of subtypes of a new ontological category of dynamic individuals called general processes" (Seibt, 2005, p. 23). A related proposal, outlining a hierarchy of such dynamic individuals by recourse to the modes of self-maintenance in various far-from-thermodynamic equilibrium systems, has been made by Richard Campbell (2009, 2015).

With regard to the dynamic construction (and co-construction) of individual human subjects and their social relations, Developmental Systems Theory, a research program initiated by Susan Oyama (2000) and now being pursued as a multi-disciplinary effort (Oyama, Gray, & Griffiths, 2001), has great promise. Organisms whose growth and reproduction require that each generation pass through the bottleneck of Weismannian segregation of germ and soma, with each individual organism, regardless of its eventual complexity, beginning its life as an undifferentiated zygote, cannot contribute anything to their single-celled embryonic offspring beyond what is contained in a small package of cytoplasm and an even smaller package of genes. Such organisms include all metazoans, and of course ourselves. The Weismannian bottleneck is too narrow for the inheritance of acquired traits. Consequently these organisms develop mechanisms for epigenetic inheritance, for passing along traits, physical scaffolding, and social structures not contained, or "coded," in genes or cytoplasm. The physical scaffolding and social structures produced and maintained by one generation condition the development and learning of members of the next, who in turn dynamically transform both scaffolding and social structures by virtue of participating in them. Individual and society co-construct each other in "cycles of contingency."

After suggesting that "Representation is the wrong unit of abstraction in building the bulkiest parts of intelligent systems" (Brooks, 1991, p. 139), Rodney Brooks set out to build embodied "creatures," validated by real time interaction with their environments, creatures unburdened by the need for detailed representational models of the world because they were equipped "to let the world itself serve as its own model." In recent years, analytic philosophy of mind has profited from the rediscovery of Merleau-Ponty and other representatives of the phenomenological tradition. Accounts of our co-constitution of and by an intersubjective world, or of the importance to human ontogenesis of what Shaun Gallagher (2009, p. 292) calls primary intersubjectivity, or what Hanne De Jaegher and Ezekiel Di Paolo (2007) have called participatory sense-making, have proliferated. For the remainder of the paper, we examine this latter proposal and the larger enactivist framework as a positive explanatory resource—highlighting some of the insights wrought from starting with sociality rather than individuality—but also as a cautionary note on what not to expect from a general account of sociality, even one based on a process approach.

4.5 Enactivism: Individuality as a Self-Maintenant Network of Processes

The enactivist literature, in which the concept of participatory sense-making originated, neglected the social dimensions of cognition for decades. Despite this lacuna, De Jaegher and Di Paolo present their concept of *participatory* sense-making as a direct extension of the enactivist notion of *sense-making*. In the present section, we briefly sketch the development of the concept of sense-making within the enactivist literature as it directly relates to and anticipates the concept of participatory sense-making.

Weber and Varela (2002) formulate the concept of sense-making in order to address what Matthews (1992) considers the central problem of modern biology—namely, giving a satisfactory account of the living individual organism. Matthews is troubled by how many theorists regard functional language within biology, despite its widespread use, as a "methodological fiction" that would be superfluous if all of an organism's physical mechanisms and evolutionary history were fully understood. Extending the earlier theory of cellular autopoeisis, Weber and Varela (2002) argue that subjectivity involves an intrinsic teleology, a "*sense-creation* purpose whence meaning comes to its [an organism's] surrounding" (p. 117; original emphasis). The basic idea is that the range of internal compensations and interactions available to an organism—which corresponds to its autopoietic self-maintenance—encompass the sense it can make of its environment. Weber and Varela thus present the concept of sense-making as an alternative to treating living functions as a methodological fiction, which as they argue entails the paradox of individuality we have discussed (Sect. 4.3). In sketching an alternative, Weber and Varela offer the beginnings of a solution to the paradox, but, as Di Paolo (2005) argues, their account of agency is still limited, treating autopoiesis as the sole foundation of agency from which sense-making emerges.

Di Paolo (2005) argues that sense-making involves two equally primitive aspects—conservation and adaptivity. To delineate these two aspects, Di Paolo uses the example of falling off of a cliff. Balancing on the edge of a cliff and falling over the edge are—as far as autopoietic self-maintenance is concerned—viable interactions, since neither causes system disintegration (p. 436). This shortsighted perspective results from the structural determinism of autopoiesis, which characterizes a system strictly in terms of its present state; hence, it is only the thud against the ground that violates conservation. This is, of course, useless to the organism that fell, since the value judgment comes only after disintegration. The limited perspective, however, doesn't indicate a fatal flaw with autopoietic enactivism, since its structural determinism underpins its systemic grounding of cognitive terminology (p. 434). Rather than jettisoning or re-defining autopoiesis, Di Paolo introduces adaptivity as a separate operational term.

While autopoietic conservation—roughly, the organism's capacity for self-preservation—provides the basic all-or-nothing norm of viable self-identity, adaptivity specifies an organism's current status in relation to its viability. Adaptivity implies that an organism possesses *some* means of anticipating and avoiding or mitigating violations of its viability boundaries. The organism attempts to mitigate if not reverse a destructive tendency by modifying an internal state or environmental coupling. Adaptivity thus entails a gradation of norms, a necessary condition for learning. Distinguishing between conservation and adaptivity, Di Paolo argues, also coheres with the two basic connotations of sense-making—namely, uncovering and generating meaning. Uncovering meaning corresponds to the basic process of conservation or self-maintenance, while generating meaning consists in retroactively attributing significance to what were previously neutral encounters based on whether the encounters caused the organism to approach towards or recede from the boundaries of viability. Di Paolo's refined concept of sense-making thereby affords him a richer account of agency.

Like the notion of priority we discussed in Sects. 4.2, 4.3, and 4.4, agency entails an asymmetry. Agential behavior is not simply an organism's structured coupling with its environment but additionally implies an organism's regulation of the coupling (Di Paolo, 2005, p. 442). The latter, unlike pure autopoietic coupling, has an intentional structure for the organism: it is possible for the organism to detect whether an action has succeeded or failed in improving its viability condition.

Di Paolo's (2009, 2010) aim is to further illustrate why cognition is not reducible to autopoietic conservation. This involves conceptualizing how the sense-making activities of an organism can radically transform its metabolic substrate (Di Paolo, 2009). Part of this conceptual groundwork consists in using the notion of operational closure as an alternative way of determining whether something is a constituent part of a living system: a process counts as part of a living system based on whether it is conditioned by another process of the system, rather than whether it is on the inside or outside of a physical boundary (Di Paolo, 2009, p. 16). Characterized in this way, it is possible to recognize cases in which an animal's behavior fundamentally reshapes the network of processes that constitute the living system and thereby becomes itself a constituent process.

Di Paolo (2010) offers the water boatman insect (*corixidae*) as a concrete example of how agency can transform an organism's metabolic substrate. By trapping air bubbles on the hairs of its abdomen, the water boatman is able to live most of its life underwater, where it feeds mostly on algae (p. 63). The insect's preference for eating algae and other underwater plants leads it to live in conditions more precarious than necessary for its survival and reproduction, a development that does not cohere with a strictly autopoietic viewpoint. Agency opens up previously non-viable domains and with it greater precariousness (p. 66). By illustrating how agency intervenes in its metabolic substrate, Di Paolo explains how cognition is intimately related with though not strictly an extension of autopoietic self-maintenance. The interrelationship between sense-making and autopoiesis bears on the dynamics between individuals and social structures, as Di Paolo (2010) explicitly notes. But a better point of entry for understanding Di Paolo and De Jaegher's enactivist approach to social interaction is to return to Di Paolo's earlier characterization of agency (Di Paolo, 2005).

The most basic way for a motile organism to regulate its coupling with its environment is via movement. The physical mechanisms that enable movement may be extremely rudimentary, such as bacteria's swimming and tumbling behavior, but even in such a simple case the organism still shapes its interactions, with some gaining positive and others negative valences. In this manner, the generative connotation of sense-making reflects a historical dimension of subjectivity. This intrinsic historicity offers a general perspective on how individuals fundamentally differ from each other, each possessing idiosyncratic interactive histories.

Yet De Jaegher and Di Paolo's formulations of *participatory* sense-making, as we detail in the next section, have focused on synchronous interactive dynamics. In particular, the authors focus on the motile aspects of social interactions. While participatory sense-making thus offers many valuable insights, there are some problematic consequences of giving general priority to the motile dimension. As becomes more apparent in De Jaegher's subsequent attempts to extend the participatory sense-making concept, this emphasis risks erasing the otherness inherent to social interaction and thereby falling prey to the same basic flaw as the social theories that they seek to supplant. Though it remains the primitive form of agency, we will argue that motility should not be given a generalized priority in analyses of social interaction.

4.6 Participatory Sense-Making: An Enactivist Approach to Sociality

De Jaegher and Di Paolo (2007) initially formulate the concept of participatory sense-making as an alternative to what they call the "Rear Window approach" to social interaction. The Rear Window approach, like the two flawed approaches to the problem of other minds we sketched in Sect. 4.3, is "cognitivist" in the sense that it treats social interaction as mediated by the communication of propositional content. In its place, in keeping with Di Paolo's aforementioned account of agency, participatory sense-making examines social interaction with a view toward how interactants ubiquitously influence each other's movements even in the absence of the more specialized mechanisms whose possibility it is meant to explain.

In two introductory articles, De Jaegher and Di Paolo (2007, 2008) focus on the two-person face-to-face encounters of everyday life. The key conceptual innovation in these articles is the idea of coordination. Two systems are *correlated* if their behavior coheres beyond what is expected of them given each system's capabilities; *coordination* simply refers to cases in which a correlation is non-accidental (De Jaegher and Di Paolo, 2007, p. 490). To counter the Rear Window approach, the authors note a wide range of coordination mechanisms including the synchronized light flashing of fireflies and the synchronized pendulum swinging of two people in view of each other (p. 490). Given the ubiquity and range of coordinating mechanisms, the authors suggest that coordinated movement is often a dynamic endogenous to interaction rather than the output of a particular cognitive mechanism. And given motility's primitive significance, such coordinated movement can play a constitutive role in many cognitive abilities.

There are four types of coordination mentioned in the two introductory articles—synchronization, mirroring, imitation, and anticipation. Synchronization receives the most attention, though all four types facilitate fluid and seamless social interaction. Although De Jaegher and Di Paolo explicitly caution that their list of coordinating factors is not exhaustive, they do not similarly qualify the significance of fluidity to interaction. The issue with generalizing the meaning of fluidity in social interaction becomes more apparent when De Jaegher attempts to apply participatory sense-making to larger social contexts, as we examine below.

At this point, we have noted two of participatory sense-making's most promising insights for social theory—namely, serving as a corrective to an exaggerated cognitivism and alleviating the pressure of anchoring cognitive abilities in individual mechanisms. While we heartily endorse these insights, we turn now to some problematic features of how De Jaegher and Di Paolo, and especially the former, have attempted to scale participatory sense-making to larger social contexts. This is not intended to be a rejection of applying the concept to larger domains but rather an argument for a more circumscribed and qualified extension. From the outset, De Jaegher and Di Paolo plan on scaling participatory sense-making in order to better inform the social sciences, laying the groundwork for this project in the introductory articles. We consider two of their efforts in this regard—their definition of what counts as social interaction and their concept of a coordination spectrum—and then show how these adversely impact the resultant framework for social theory, as articulated by De Jaegher. Hence, for the remainder of the section, we critically scrutinize the development of participatory sense-making beyond two-person face-to-face interaction.

One objection to participatory sense-making that De Jaegher and Di Paolo anticipate is that the wide range of coordination factors makes social interaction indistinguishable from other kinds of coupling. Distinguishing *social* interaction from non-social cases is important for delimiting a domain of research. Towards this end, De Jaegher and Di Paolo (2008) define social interaction as:

> The regulated coupling between at least two autonomous agents, where the regulation is aimed at aspects of the coupling itself so that it constitutes an emergent autonomous organization in the domain of relational dynamics, without destroying in the process the autonomy of the agents involved (though the latter's scope may be augmented or reduced) (p. 39).

Autonomy, in the context of enactivism, refers to a system's ability to sustain itself under precarious conditions. Organisms are such systems and, De Jaegher and Di Paolo suggest, so are social interactions: unlike many other kinds of interactions among people, *social* interactions gain a momentum of their own with properties endogenous to the interaction that help sustain it beyond the intentions of any one individual. The definition thus excludes cases in which one individual lacks any control over a situation, such as a hostage scenario. One of De Jaegher and Di Paolo's aims is to show that it is possible for this strictly operational definition to recapture our intuitive notion of sociality. Yet the shortcomings of giving a strictly operational definition are already hinted at with the qualification that social interaction "may augment or reduce" the autonomy of the interactants. The issue is one of interpreting autonomy as augmented or reduced along a single axis, and signals the guiding theme of De Jaegher's expansion of participatory sense-making to larger social contexts.

De Jaegher's efforts center on the spectrum of social coordination, the second piece of groundwork for scaling participatory sense-making that appears in the introductory articles. This spectrum ranges over degrees of coordination: on one side of the spectrum lies mere "orientation" and, on the other, "joint sense-making." In the latter case, it may be very difficult or even impossible to tease out each individual's distinct contributions. On the orientation end of the spectrum, suppose I happen upon a person looking for something and point a finger in order to direct his or her search (De Jaegher & Di Paolo, 2008, p. 42). In this case, it is quite easy to delineate each individual's contributions: one is the searcher and the other a helpful pointer; importantly, the searcher's desire would exist irrespective of my presence. For interactions more properly described as involving "joint sense-making", it may be very difficult or even impossible to tease out each individual's distinct contributions. The social coordination spectrum thus serves as a useful tool in assessing the interdependence of an interaction.

De Jaegher, Di Paolo, and Gallagher (2010) tentatively propose the coordination spectrum as a general indicator of the affective character of social interaction. The proposal has some empirical support within the psychotherapeutic context, where there is a correlation between, on the one hand, high bodily coordination of therapist and patient and, on the other, a patient's positive assessment of the therapy session (p. 442). Over a series of articles, De Jaegher expands on the social coordination spectrum, with emphasis on the suggestion that high coordination correlates with positive affective states. While the concept of the social coordination spectrum is useful, as is the above definition of social interaction, we take issue with how they have been extended. Most importantly, even though the connection between the coordination spectrum and the affective character of interaction is indexed to a specific context in De Jaegher et al. (2010), this qualification and any other proviso is scanty if not entirely absent from De Jaegher's subsequent efforts.

De Jaegher proposes a number of related spectra. Among her many efforts, we will concentrate on McGann and De Jaegher (2009) and De Jaegher (2013): the former captures the tight link between the spectra and the privileging of motility, while the latter shows the extent to which De Jaegher is willing to give motility explanatory priority. The main goal of McGann and De Jaegher (2009) is to give an account of social perception in terms of social contingencies and thereby link participatory sense-making to O'Regan and Noë's (2001) sensorimotor contingency theory. Towards this end, McGann and De Jaegher recast joint sense-making and orientation, the opposite poles of the social coordination spectrum, in terms of fluid and rigid interactions, respectively. In this context, fluidity appears as sensitivity to social contingencies. That is to say, being socially skilled entails the ability to improvise based on our own and others' changing emotional states (p. 427–428). Lack of social skill, by contrast, implies rigidity and predictability, an inability to engage ones particular audience. In making this connection with O'Regan and Noë's sensorimotor contingency theory, McGann and De Jaegher erode some of participatory sense-making's value: interpreting sociality in terms of unpredictability dovetails with O'Regan and Noë's general privileging of quantifiable movement over first-personal experience. Likewise, De Jaegher's application of fluidity and rigidity to political theory suggests that she too believes that motility holds a general explanatory priority over first-personal experience.

Drawing upon Gilligan and Richards (2009), De Jaegher (2013) frames the fluid-rigid dichotomy in terms of horizontal democratic forms of association in contrast to top-down patriarchal social organizations, respectively. The connection is ambitious, representing an attempt to show participatory sense-making's relevance to analyzing entire societies. This extension is predicated on yet another spectrum, the spectrum of power symmetry. De Jaegher writes, "When interacting with another person, or with an institution, one or other partner may be more or less 'dominant,' more or less influential. In such asymmetric relations or interactions, the degree of influence that each partner has is different" (p. 23). De Jaegher's understanding of dominance as a matter of "more or less" links the previous notion of autonomy to the social coordination spectrum: on the less social end, highly asymmetric institutionalized interaction corresponds to orientation, since it is easy, so the logic goes, to delineate the differing contributions by reference to dominance relations. On the other end, De Jaegher offers the example of a lawyer-client relationship as a case of greater symmetry/joint sense-making: the lawyer and client work together to interpret the law to their advantage. One immediate issue with this generalized use of the fluid-rigid dichotomy is illustrated by a closer examination of one of the studies that De Jaegher cites as support.

Isabela Granic (2002) uses dynamic systems theory to model parent-child relationships. Granic is specifically interested in how patterns of hostility between parents and children can emerge despite each party's amicable intentions, a phenomenon she attributes to behavioral attractors (p. 273). In both McGann and De Jaegher (2009) and De Jaegher (2013), Granic's cases of hostile parent-child behavioral patterns are glossed as a rigid pattern of social interaction: the hostility occurs *contrary* to each individual's intentions, thereby reducing the autonomy of each, and so constitutes a rigid pattern in the sense that it predictably occurs irrespective of the specific circumstances of the encounter. But Granic herself does not think of hostile and other negatively valenced interactions as being more "rigid" than positive ones. Behavioral attractors *stabilize* both positively and negatively valenced patterns of interactions. From a therapeutic perspective, Granic is most interested in identifying and figuring out how to destabilize negative patterns, but this doesn't imply that negative patterns of interaction possess any greater stability or "rigidity" than positive ones.

Even if movement is the most primitive form of agency, as Di Paolo asserts, this does not translate into a general explanatory priority. This follows from two other parts of Di Paolo's (2005, 2009, 2010) account of agency: the inherently historical dimension of subjectivity and, second, the capability for subjectivity to intervene in and transform its physical substrate. Both of these features are highlighted in Wrangham's cooking hypothesis, discussed in Sect. 4.2. Not only does the cooking hypothesis suggest that elements of our gross physical anatomy are evolutionary products of social structures but, more directly related to the present discussion, that cooking facilitated an increase in individuals' mobility due to nutritional gains that in turn enabled the shrinkage of hominin intestines (Wrangham, 2009, p. 44).

The cooking hypothesis indicates a social dimension that can only be captured diachronically, in this case on a fairly distal evolutionary timescale. We are not suggesting that such timescales must always be taken into consideration when examining social interaction, but rather arguing against treating movement as a discrete, isolable dimension of social interaction, as implied by De Jaegher's one-dimensional spectra of sociality and autonomy.

In addition to neglecting historical contexts, the generalization of the fluid-rigid dichotomy takes priority over first-personal experience. This is evident in De Jaegher's identification of patriarchy with rigid social interactions. Underpinning the proposal is an equivocation of rigidity with rule-governed behavior. Hierarchical social structures are presented as inherently more rigid than horizontal ones. But the issue with this proposal is that there is not an intrinsic link between, on the one hand, rigid social interaction and, on the other, rules or hierarchies. Both rules and hierarchies can enable novel and desirable cognitive abilities. Take for instance the profession of modern science: it is doubtful whether it could exist without the systems of credit and the hierarchical institutions implied therein (Kitcher, 1990). To the extent to which there is a morally relevant difference between the hierarchies of scientific institutions and patriarchal institutions, it should not be understood in terms of the respective motility of each institution's members; this would be to put the cart before the horse, so to speak. The difference should instead be understood in terms of members' first-personal experiences: with regard to these two cases, the morally salient difference concerns the suffering of those subjected to oppression under patriarchal institutions. That De Jaegher gives priority to the motility profile reflects less a considered judgment than her ambition to provide a research framework for the social sciences.

Though De Jaegher's research framework possesses some notable merits, it comes at a steep price. By obscuring first-personal experience and the historical dimensions of subjectivity—our two main reservations with De Jaegher's extension of participatory sense-making—the research framework effectively dilutes if not erases the encounter with otherness inherent to social interaction, just as the two approaches to the problem of other minds canvassed in Sect. 4.2 erase rather than address it. In giving motility a generalized priority, sociality is reduced to a one-dimensional spectrum and anything that cannot be mapped onto the spectrum, including most of each individual's idiosyncrasies and personal history, is set aside. In this regard, De Jaegher's development runs the risk of exchanging the exaggerated cognitivism of the Rear Window approach for a flattened notion of agency that collapses into a small number of spectra sharing no common metric. The impetus behind De Jaegher's various spectra of sociality is to provide a quantifiable measure of social interaction and thereby a tractable object of empirical research. In our closing section, we briefly sketch how to recontextualize De Jaegher's framework in order to avoid obscuring the otherness of social interaction.

4.7 Conclusion: Scaling Participatory Sense-Making Without Erasing Otherness

Process-based approaches in the social sciences hold great promise and have already brought a number of advances. In the first half of this paper, we focused on the need to overcome the POI thesis and the attendant paradox of individuality. De Jaegher and Di Paolo's concept of participatory sense-making and the enactivist movement of which it is a part achieve both of these goals. Yet despite these and other insights, the concept of participatory sense-making and process-based approaches more generally remain vulnerable to some of the same issues found in the social theories that they are supplanting.

In the second half, we argued that recognizing the alterity inherent to social interaction is a crucial standard for assessing social theories. Although we first broached the issue of otherness in connection with the POI thesis, De Jaegher's application of participatory sense-making to larger social contexts has problematic implications as well. By interpreting the motility profile of a social interaction as possessing a meaning independent of any of the interaction's historical or first-personal dimensions, the idiosyncrasy of the interactants is obscured. The result is a hollow collectivism, which serves as the converse error to the atomic individualism of the POI thesis. Fortunately, overcoming the problematic features of De Jaegher's framework may consist simply in qualifying its claims.

One of De Jaegher's primary aims is to demonstrate that a process-based approach offers a suitable framework for quantified measures of social interaction and thereby for controlled experiments. This is a crucial argumentative move, as it suggests that process approaches can fully supplant substance-based social theories. Yet as the alterity issue implies, quantifiable measures and controlled experiments do not exhaust the research methodology of the social sciences, nor should quantified measures be understood in strict isolation from historical and phenomenological considerations. Operational terminology is extremely valuable, as evidenced by Di Paolo's (2005) account of agency, yet sociality is an instance of a domain that cannot be exclusively explained in its terms. It is important not to confuse the rhetorical move of showing how process-based approaches translate into fruitful experimental research programs with a mature process approach to social interaction. We hope to have gone some way toward motivating research into the prospect that, in a process framework, such interaction might best be thought of as prior to both individuality and sociality.

Our sketch of an individuality without atomism and a sociality without collectivism suggests that a process-based framework for social theory will be deeply interdisciplinary, with the methodologies of both experimental and historical sciences (in the sense of Cleland, 2002) at its heart. A process-based framework for social robotics will likewise be interdisciplinary. It will have to confront the problem of alterity from the outset, conceiving of the social robot as encountering and interacting with a genuine *other*, and not merely with another instance of the same, familiar agency template it itself implements. Such a being will value the other as *other*, as an entity not merely numerically distinct from itself. This remains, however, a difficult challenge for humans in their dealings with fellow humans. Robot agents compound the challenge—whether on the battlefield, in search and rescue, or in the hospital. It would be naïve to expect them to short-circuit it.

References

Brandom, R. (1994). *Making it explicit*. Cambridge, MA: Harvard University Press.

Brooks, R. (1991). Intelligence without representation. *Artificial Intelligence, 47*, 139–159.

Campbell, R. J. (2009). A process-based model for an interactive ontology. *Synthese, 166*, 453–477.

Campbell, R. J. (2015). *The metaphysics of emergence*. London: Palgrave.

Churchland, P. M. (2013). *Matter and consciousness*. Cambridge, MA: MIT Press.

Cleland, C. E. (2002). Methodological and epistemic differences between historical science and experimental science. *Philosophy of Science, 69*, 447–451.

Davidson, D. (1984). *Inquiries into truth and interpretation*. Oxford: Oxford University Press.

De Jaegher, H. (2013). Rigid and fluid interactions with institutions. *Cognitive Systems Research, 25*, 19–25.

De Jaegher, H., & Di Paolo, E. (2007). Participatory sense-making: An enactive approach to social cognition. *Phenomenology and the Cognitive Sciences, 6*, 485–507.

De Jaegher, H., & Di Paolo, E. (2008). Making sense in participation: An enactive approach to social cognition. In F. Morganti, A. Carassa, & G. Riva (Eds.), *Enacting intersubjectivity: A cognitive and social perspective on the study of interactions* (Vol. 10, pp. 33–47). Amsterdam/Washington, DC: IOS Press.

De Jaegher, H., Di Paolo, E., & Gallagher, S. (2010). Can social interaction constitute social cognition? *Trends in Cognitive Sciences, 14*, 441–447.

Di Paolo, E. (2005). Autopoiesis, adaptivity, teleology, agency. *Phenomenology and the Cognitive Sciences, 4*, 429–452.

Di Paolo, E. (2009). Extended life. *Topoi, 28*, 9–21.

Di Paolo, E. (2010). Overcoming autopoiesis: An enactive detour on the way from life to society. *Advanced Series in Management, 6*, 43–68.

Dummett, M. (1981). *The interpretation of Frege's philosophy*. Cambridge, MA: Harvard University Press.

Fodor, J. A. (1975). *The language of thought*. Cambridge, MA: Harvard University Press.

Fodor, J. A. (1981). The present status of the innateness controversy. In *Representations: Philosophical essays on the foundations of cognitive science* (pp. 257–316). Cambridge, MA: MIT Press.

Gallagher, S. (2009). Two problems of intersubjectivity. *Journal of Consciousness Studies, 16*, 289–308.

Gilligan, C., & Richards, D. A. (2009). *The deepening darkness: Patriarchy, resistance, and democracy's future*. New York: Cambridge University Press.

Granic, I. (2002). The self-organization of parent-child relations: Beyond bidirectional models. In M. D. Lewis & I. Granic (Eds.), *Emotion, development, and self-organization: Dynamic systems approaches to emotional development* (pp. 267–297). New York: Cambridge University Press.

Hull, D. (1987). Geneaological actors in ecological plays. *Biology and Philosophy, 1*, 44–60.

Kant, I. (2012). *Groundwork of the metaphysics of morals*. Cambridge, MA: Cambridge University Press. (Originally published in 1785).

Kitcher, P. (1984). *The nature of mathematical knowledge*. Oxford: Oxford University Press.

Kitcher, P. (1990). The division of cognitive labor. *The Journal of Philosophy, 87*, 5–22.

Levine, A. (2001). Individualism, type specimens, and the scrutability of species membership. *Biology and Philosophy, 16*, 325–338.
Mann, W.-R. (2000). *The discovery of things*. Princeton: Princeton University Press.
Matthews, G. (1992). De anima 2.2-4 and the meaning of life. In M. Nussbaum & O. Rorty (Eds.), *Essays on Aristotle's de anima* (pp. 185–193). Oxford: Clarendon.
McGann, M., & De Jaegher, H. (2009). Self-other contingencies: Enacting social perception. *Phenomenology and the Cognitive Sciences, 8*, 417–437.
Merleau-Ponty, M. (2012). *The phenomenology of perception*. New York: Routledge.
Mill, J. S. (2002). *Utilitarianism*. Indianapolis: Hackett. (Originally published in 1863).
O'Regan, J. K., & Noë, A. (2001). A sensorimotor account of vision and visual consciousness. *Behavioral and Brain Sciences, 24*, 939–973.
Oyama, S. (2000). *Evolution's eye*. Durham: Duke University Press.
Oyama, S., Gray, R. D., & Griffiths, P. E. (2001). *Cycles of contingency: Developmental systems and evolution*. Cambridge, MA: MIT Press.
Seibt, J. (2000). Pure processes and projective metaphysics. *Philosophical Studies, 101*, 253–289.
Seibt, J. (2003). *Process theories: Crossdisciplinary studies in dynamic categories*. Dordrecht: Kluwer.
Seibt, J. (2005). Process ontology. In G. Imaguire (Ed.), *Metafisica e ontologia*. München: Philosophia Verlag.
Sellars, W. (1956). Empiricism and the philosophy of mind. In H. Feigl & M. Scriven (Eds.), *The foundations of science and the concepts of psychology and psychoanalysis* (Minnesota studies in the philosophy of science, Vol. I, pp. 253–329). Minneapolis: University of Minnesota Press.
Sellars, W. (1960). Being and being known. *Proceedings of the American Catholic Philosophical Association, 34*, 28–49.
Weber, A., & Varela, F. J. (2002). Life after Kant: Natural purposes and the autopoietic foundations of biological individuality. *Phenomenology and the Cognitive Sciences, 1*, 97–125.
Wrangham, R. (2009). *Catching fire: How cooking made us human*. New York: Basic Books.

Chapter 5
Social Meta-Learning: Learning How to Make Use of Others as a Resource for Further Learning

Jedediah W.P. Allen and Hande Ilgaz

Abstract While there is general consensus that robust forms of social learning enable the possibility of human cultural evolution, the specific nature, origins, and development of such learning mechanisms remains an open issue. The current paper offers an action-based approach to the study of social learning in general and imitation learning in particular. From this action-based perspective, imitation itself undergoes learning and development and is modeled as an instance of social meta-learning – children learning how to use others as a resource for further learning. This social meta-learning perspective is then applied empirically to an ongoing debate about the reason children imitate causally unnecessary actions while learning about a new artifact (i.e., over-imitate). Results suggest that children over-imitate because it is the nature of learning about social realities in which cultural artifacts are a central aspect.

Keywords Social learning • Over-imitation • Cultural affordances • Self-scaffolding • Social realities

5.1 Theoretical Framework

Neither humans nor robots can be preprogrammed to competently operate in real world physical environments: there is too much *complexity* and it cannot be anticipated beforehand what will be *relevant* for successful interaction (McCarthy & Hayes, 1969). Instead, both humans and robots must be capable of learning. In the realm of social engagements, the issue of preprogramming is even more pronounced because the "objects" involved are ontologically emergent from the co-constituted activity of the participants (Bickhard, 2008). Accordingly, participation in a socio-cultural world requires that humans and robots be equipped with the ability to learn

J.W.P. Allen (✉) • H. Ilgaz
Psychology Department, Bilkent University, 06800, Ankara, Turkey
e-mail: jallen@bilkent.edu.tr; jedediahwpallen@gmail.com; hande.ilgaz@bilkent.edu.tr

R. Hakli, J. Seibt (eds.), *Sociality and Normativity for Robots*, Studies in the Philosophy of Sociality 9, DOI 10.1007/978-3-319-53133-5_5

from others (social learning), and, for human-like competence, to learn and develop their ability to learn from others (social meta-learning).

There is general consensus that participation in human culture and advances through cultural evolution have depended on the possibility of robust forms of social learning. Imitation is widely thought to be one of these robust forms of social learning but the specific nature, origins, and development of this form of learning garners far less agreement (Anisfeld, 2005; Jones, 2007; Gergely & Csibra, 2005; Heyes, 2005; Tomasello, 1996). Further, the dominant theoretical frameworks in both animal and developmental research do not motivate a perspective in which imitation itself involves learning and development. In general, humans are not only capable of learning, but also, they are capable of learning to learn (meta-learning, Bickhard, 2005). With respect to social forms of meta-learning children are learning how to make use of others as a resource for subsequent (social/individual) learning and development. This perspective on imitation, as a form of social meta-learning, is motivated from within an alternative action-based framework for modeling development more broadly (Allen & Bickhard, 2013a; Bickhard, 2009).

5.1.1 An Action-Based Framework: Interactivism

Theoretical models and assumptions about the nature and origins of representation, learning, and development involve sets of mutually constraining relations (Allen & Bickhard, 2011a; Piaget, 1954). Using an action-based framework for modeling learning and development means being committed to (inter)action as the epistemic ground for the emergence of mind. For an action-based perspective, knowledge is *constituted* by interactive competence and representation is a matter of learning about relevant interactive possibilities.

Interactivism is a well-developed action-based framework that models the origins and ontology of representation, learning, and development for both the physical and social world (Bickhard, 1980, 2009, in preparation; Bickhard & Terveen, 1995). The interactivist model of representation is ultimately grounded in the normative functioning of recursively self-maintaining systems that are far from thermodynamic equilibrium. Specifically, the minimal model of representation explicates how *truth-value* and *aboutness* are considered emergent properties of systems capable of successfully anticipating interaction potentialities that would be functional within the current situation (Bickhard, 2009). In this model, anticipation forms the core of representation.

5.1.1.1 Object Representation

The emergence of a more canonical form of representation has been demonstrated by borrowing from Piaget's own action-based model of object representation (Piaget, 1954). For interactivism, object representation is constituted by an invariant web of anticipations for potential interaction. Any point in this web of anticipations is reachable from any other point and the web remains invariant under a large class of other possible transformations (Bickhard, 2009). For example, an infant's interactive representation of a small wooden block is constituted by various possible visual scans, finger manipulations, and mouthing activities. These interactive possibilities remain available despite multiple other transformations such as visible and invisible displacement, occlusion, containment, and so on; however, if the block involves a transformation in which it is burned or crushed, the previously available interactive possibilities are gone.

5.1.1.2 Social Agent Representation

In contrast to physical objects, the interactive possibilities afforded by social agents are, largely, not perceptually available. Instead, your interactive characterization of a social agent depends on that other agent's interactive characterization of you. This creates an unbounded iteration of mutually dependent characterizations that must get resolved in order for the agents to successfully interact. Consequently, an accurate interactive characterization of a social agent is going to require knowing about the broader type of situation of which the two agents are major aspects. Therefore, successful interaction and coordination between social agents is made possible through the creation, invocation, and maintenance of mutually held interactive characterizations of the situation – i.e., social realities. In turn, social realities provide the basic ontology for learning about and navigating social life (Bickhard, 2008).

5.1.1.3 Learning and Development

From an action-based perspective, knowledge is constituted by interactive competence and, is therefore, inherently active, relational, and necessarily constructed through a variation and selection process (Bickhard, 2009). In contrast to nativism and empiricism, knowledge is not assumed to come from somewhere (genes or environment), nor is it assumed to be constituted by some sort of correspondence relationship between the mind and the world (Allen & Bickhard, 2013b; Bickhard, 2009). Further, for an action-based approach, learning is not assumed to require a base set of representational primitives (Allen & Bickhard, 2011b). Instead, knowledge is actively constructed through the creation of new internal organizations for the organism's possible functioning and internal organizations that enable successful interaction with the environment will be selected for. In short, learning is modeled as a variation and selection, emergent-constructivist, process (Allen & Bickhard, 2011a).

For organisms that are capable of minimal forms of development, the constructive processes can be recursive in that current constructive processes can make use of prior constructive processes.[1] In more complex organism, like humans, the constructive processes can themselves undergo learning. Such organisms will be capable of meta-recursive constructive processes – they will be cable of learning to learn (Bickhard, in preparation). From this perspective learning and development involve the same underlying dynamics but at different times scales. Accordingly, development will be constituted as the properties and constraints manifested by the historicities of learning – the ways in which prior learning influences future learning (Campbell & Bickhard, 1992). For developmental research this will mean designing studies and interpreting results in the developmental context of what children have previously learned about the affordances of objects and social agents in different types of situations.

5.1.2 Social Meta-Learning

With respect to social forms of meta-learning children are learning how to make use of others as a resource for subsequent (social/individual) learning and development. In general, forms of social meta-learning are going to involve issues of the competence and the reliability of those others (Bickhard, in preparation). An early example of social meta-learning is when children develop attachment relationships that are adapted to specific people in their care-giving environment. Different attachment relationships emerge to enable children to maximize their use of the adult as an emotional resource. Children who develop a secure attachment relationship are able to consistently use the adult caregiver to help make *global* evaluations about situations and in that sense use the adult as a secure base from which to explore their environments. A later emerging form of social meta-learning is social referencing. Social referencing involves using an adult's emotional reaction to a particular object or event within a specific situation. This emotional reaction provides the child with a *more specific* evaluation about whether to proceed or withdraw from their ongoing activity.

Imitation is a powerful form of social meta-learning that remains particularly useful throughout the first 3–4 years of life. For imitation, children are learning about how to make use of adults as a resource for specific engagements with the physical, social, and cultural world. However, children must not only learn *how* to imitate another person's actions (solving the correspondence problem – Nehaniv & Dautenhahn, 2002), but also: *who* to imitate, *what* to imitate, and *when* to imitate. Further, children must develop the ability to use imitation for different functions: learning, play, affiliation, trust, commitment, etc.

[1]Nativism and empiricism also have forms of recursive learning but both are restricted to the combinatorial spaces defined by their base sets of representational primitives and do not allow for emergent representation (Allen & Bickhard, 2011a).

Finally, social meta-learning has renewed importance once children develop sufficient language capabilities. "Trust" research has focused on the various cues that children learn to use as indicating the reliability of information coming from conflicting sources (Harris et al., 2012). In general, many of these cues are learned and reconciled with other cues at around age four. For example, at around age four, a person's past accuracy is recognized as a more relevant cue for predicting their future reliability over an affiliation cue like familiarity (Corriveau & Harris, 2009).

Having outlined an action-based approach to the representation of social realities and the development of social meta-learning, we will apply this perspective to the empirical study of imitation learning. This application has two aspects: first, it proposes to replace the current focus on imitation fidelity with a perspective in which all imitation is selective and based on cognitive, motivational, and social-cognitive processes. Second, it proposes an alternative account for an ongoing debate within contemporary imitation research about the nature of over-imitation. The paper will conclude by suggesting that the proposed action-based framework is adequate to the task of grounding empirical research on imitation in a way that more closely resembles the richness of human culture and provides greater theoretical integration with other developmental phenomena as compared to extant frameworks. Such theoretical integration is crucial for robotics programs that want to develop systems capable of participating in the social realities of human culture. Further, only action-based frameworks have been able to naturalize representation and mind, at least at the level of theoretical explanation. In turn, such naturalism is a perquisite for the possible construction of artificial systems with the various emergent properties of relevance for developing social ontology.

5.2 The Current Study

5.2.1 The Debate: Selective vs. Over-imitation

Differences in the fidelity with which young children copy an adult demonstration has formed the basis for multiple debates and controversies regarding imitation learning in particular and social cognition more broadly (Call & Carpenter, 2002; Lyons, 2009; Meltzoff, 1995; Whiten, 2006; Zentall, 2006). Demonstrations of low fidelity copying have often led to characterizations of imitation as being selective, flexible, and rationally applied to the specifics of a situation (Gergely, Bekkering, & Kiraly, 2002; Southgate, Chevallier, & Csibra, 2009; Williamson & Markman, 2006). In contrast, demonstrations of high fidelity copying have led to characterizations of imitation as being direct, automatic, and tied by evolutionary selection pressures to the intentional actions of an adult model (Horner & Whiten, 2005; Lyons, Young, & Keil, 2007; McGuigan, Whiten, Flynn, & Horner, 2007). These contrasting characterizations between selective and non-selective imitation have become particularly salient given several powerful demonstrations that young children will replicate causally unnecessary transformations on artifacts despite *direct* evidence that they are unnecessary. For example, children will replicate tapping on the top of a box as part of a sequence along with other causally necessary steps in their effort to open it. This phenomenon has been termed "over-imitation".

The current paper will address two intimately related issues: first, how to reconcile the conflicting characterizations between selective imitation and automatic over-imitation? Second, how to account for why children over-imitate? It is important to highlight that selective and over-imitation tend to manifest in different types of situations and at different ages. Specifically, selective imitation has tended to be demonstrated mostly during the second year of life and in situations that concern the selection of which *actions* to use to achieve some outcome on an object (e.g., selecting to use your head versus your hand to push a button to turn on a light). In contrast, over-imitation[2] has been demonstrated most clearly after age three and in situations that concern the selection of which *object-transformations* to use without regard for the precise actions involved (e.g., removing a horizontal pin from a box without regard for whether the pin is pushed through or pulled out). Further, when the same task has been used in different studies with both younger and older age groups, there is a transition from more to less selectivity (Gergely, 2003; Huang, Heyes, & Charman, 2006; McGuigan & Whiten, 2009). The "developmental inversion" from more to less selectivity is difficult to reconcile if researchers are focused exclusively on *cognitive* aspects of children learning about *objects* because older children should be cognitively more competent about the working of artifacts not less. In short, why do more knowledgeable older children imitate object transformations that are "clearly" irrelevant while their less knowledgeable younger peers show more selectivity? This problem is further illustrated by the demonstration that adults over-imitate with even greater fidelity than young children (McGuigan, Makinson & Whiten, 2011).

5.2.2 *Resolving the Debate: Cultural Affordances*

Uzgiris (1981) proposed a developmental transition in children's use of imitation that could provide a potential answer to the current developmental question. Uzgiris argued that the function of imitation transitions developmentally from serving a cognitive learning function to serving a predominantly social engagement function. Applied to the current debate, when young children display *selectivity* while learning something new in an imitation situation, they are manifesting aspects of their *cognitive* understanding. Further, when older children over-imitate, they

[2]Over-imitation is not just high fidelity imitation but also involves "clear" evidence that the high level of fidelity is causally unnecessary (Lyons et al., 2007).

are manifesting an underlying *social* motivation for social engagement (Call & Carpenter, 2009; Nielsen, 2006; Nielsen & Blank, 2011; Nielsen, Simcock & Jenkins, 2008).

In contrast to the social-motivational explanation, other researchers have argued that over-imitation manifests as a by-product of a useful cognitive-learning strategy in which copying all of the adults' intentional actions is typically adaptive (Lyons, 2009; Lyons, Damrosch, Lin, Deanna & Keil, 2011; Lyons et al., 2007; Whiten, McGuigan, Marshall-Pescini, & Hopper, 2009). These researchers assume that their experimental situations primarily involve an *instrumental* motivation to achieve a demonstrated outcome (e.g., to retrieve the contents in a novel container).[3] Therefore, they interpret the imitation of causally irrelevant actions as a consequence of changes in the child's cognitive understanding about how an artifact actually functions. Accordingly, these researchers conclude that when children imitate what seem to be obviously irrelevant actions (e.g., tapping on the top of a container), it is because they believe that these steps are causally necessary for achieving the ultimate goal of retrieving the contents from within the apparatus.

From the current perspective, we agree with the social-motivational approach that children are motivated to interact socially with adults; however, success at such an interaction requires learning. That is, in order to successfully interact with adults, children must learn about the interactive affordances of the situation in which artifacts are going to be a central aspect. Therefore, we also agree with the cognitive-learning approach that children are learning about the artifact; however, the meaning of what they are learning is intrinsically tied to the broader social interaction with the adult. Consequently, children's imitative responses will have less to do with their understanding of the causal necessity of different object transformations and more to do with the *culturally relevant* object affordances that children presume to be available for successful interaction in the situation.

For an example, consider learning how to play billiards through imitation. A model demonstrates how the pool cue is used to hit the cue ball and knock other balls into the holes. If an individual were to use all of his knowledge about causal necessity he would conclude that it would be more efficient to use his hands rather than the pool cue. After all, the goal is to get the balls into the holes. However, we would not consider this selective imitation (or even emulation). Instead, we would interpret this behavior as an indication that this individual is unwilling to participate in the social reality we call "billiards". In general, social realities are constituted by individuals who possess a shared interactive characterization of the situation (Bickhard, 2008). Learning about these shared characterizations suggests a sense in which imitation is able to contribute to the sharing of culture more broadly – through the learning of new social realities. Importantly, imitation situations are

[3]Lyons and colleagues (2007, 2011) have attempted to specifically test this assumption with different conditions from two sets of studies as part of their effort to rule out the alternative social motivational interpretation for over-imitation in general; however, the current study suggests that these alternative control conditions have been inadequate.

somewhat unique in that they also constitute a class of social reality themselves. Thus, as a social reality, imitation involves learning and development, while also being used to teach children about new social realities. Consequently, imitation serves as a powerful social bootstrapping mechanism for learning about culturally relevant social realities.

The above example serves to illustrate the basic point about the interactive nature of social realities involving cultural artifacts; however, it also highlights the limits of trying to define different forms of social learning solely in terms of the fidelity of the match between model and imitator. The very distinction between low fidelity selective imitation and high fidelity over-imitation derives from a theoretical perspective in which imitation is defined by a somewhat arbitrary degree of abstraction – how perceptually similar must the child's actions be in order to qualify as selective or not (McGuigan & Whiten, 2009)? Imitation always involves some degree of abstraction away from the modeled demonstration,[4] therefore, imitation always involves some degree of selectivity about what does and does not get copied. From the perspective being advocated in the current study, all imitation activity is selective and that selectivity always involves both cognitive and motivational aspects that are themselves relative to the child's understanding of the broader type of situation.

5.2.3 Imitation as a Type of Social Situation

To date, little of the research on over-imitation has directly examined the influence of the broader type of situation on the fidelity with which preschool children copy an adult model. Part of the power of the original study by Gergely et al. (2002) was to highlight the essential role that children's understanding of the broader type of situation has on how they interpret the activity from an adult demonstration. Despite some controversy regarding particular findings (Paulus, Hunnius, Vissers, & Bekkering, 2011), one of the central lessons from the research program of Gergely and his colleagues is that children's selection of what to imitate depends on what they presume to be new or relevant about the adult demonstration (Csibra & Gergely, 2009; Kiraly, Csibra, & Gergely, 2013). While these presumptions of relevance may improve and change with the development of children's learning about artifacts and agents ("rational" constraints being a part of this learning), they will always be relative to what the child understands about the broader type of social situation as well.

In general, children have an extensive developmental history of experiences interacting with different types of objects and with adults in different types of situations. Accordingly, they have learned to apply different sets of general

[4]At bare minimum, it requires abstraction away from the actions of the model to those of the imitator.

expectations about how to interact depending on the type of situation that has been indicated (pedagogical, imitation, play, helping, etc). For imitation, these *general* expectations involve making a variety of presumptions about what aspects of the *current* demonstration are potentially relevant. These presumptions are often tentative and always fallible but children have learned that some things are much more likely than others in any given situation (Schulz, Hooppell, & Jenkins, 2008). Further, what is potentially relevant in a given situation depends on one's purposes.

Children come to learn that there are different types of imitation situations that may emphasize the different functions that imitation can serve. The two most common functions in the literature concern the "cognitive-learning" and "social-motivational" functions. However, the cognitive-learning function can be extended to learning about *cultural* artifacts in the context of learning about social realities. Learning about social realities constitutes the more general phenomenon that we argue is captured by standard over-imitation studies. Therefore, the current study sought to explore how the presumptions that children make during an adult demonstration of operating a novel artifact depends on their previously learned knowledge about artifacts as physical/cultural objects (cognitive aspects), current purposes (motivational aspects), and how both of them are relative to the child's knowledge about the broader type of social situation (social-cognitive aspects).

The general procedure for the current study was as follows: During the demonstration phase of the experiment (i.e., the imitation situation), all children were given four demonstration and response trials for how to open two identical boxes. The demonstration involved five distinct actions but only two of them were causally relevant to open the box. The demonstration phase was followed by the cooperation phase (i.e., the helping situation) in which children were given the opportunity to use what they had learned about opening the boxes to help an adult confederate find a toy that was hidden inside one of them. However, the boxes were surreptitiously locked and would only open if all five target-actions were performed.

5.2.4 *Rationale for the Hypotheses*

5.2.4.1 Hypothesis 1

Although a developmental trend from more to less selective imitation seems clear when looking across the literature as a whole, there is little *direct* evidence for a specific developmental increase in over-imitation for children in the preschool years (Lyons et al., 2007; McGuigan et al., 2007). In the current study, it was hypothesized that over-imitation would increase in age such that 3-year-olds would copy irrelevant actions the least, followed by 4-year-olds, with 5-year-olds copying irrelevant actions the most.

5.2.4.2 Hypothesis 2

Similar to the general paradigms of other over-imitation research (Horner & Whiten, 2005; Lyons et al., 2007; McGuigan et al., 2007), children were invited to learn about an unfamiliar artifact through adult modeling. Specifically, the adult experimenter demonstrated a sequence of five target actions with a wooden box that resulted in the box being opened. Only two of these interactions were causally relevant to open the box and the other three were increasingly less plausible as being potentially relevant (i.e., rotating a non-functional lever, pushing a large button, tapping on the box three times). It was hypothesized that children would display differential copying of the five target actions in accordance with their prior experience such that the two causally relevant steps (removing the pin & pulling the knob) were copied most.

5.2.4.3 Hypothesis 3

The cognitive-learning perspective on over-imitation assumes that children imitate the demonstrator with an *instrumental* motivation to open a novel artifact (Horner & Whiten, 2005; Lyons et al., 2007, 2011). Thus, the steps children choose to copy are assumed to be instrumental for the goal of opening that artifact; therefore, the steps that they choose are interpreted as an indication of their understanding of how the artifact actually functions. However, the current study challenges the claim that canonical over-imitation situations primarily involve an instrumental motivation and, thus, the conclusion that children's copying behavior is an indication of what they understand to be causally necessary about opening the artifact.

Instead, because children have extensive experience with imitation type situations, they have a set of expectations about how to interact with the adult and are motivated to copy the adult's actions as much as they are motivated to open the artifact. Therefore, the current study compared children's actions on the box during a canonical over-imitation situation with their actions during a helping situation. The idea here is that a helping situation primarily involves an instrumental motivation because opening the box becomes decidedly instrumental for the broader goal of helping the adult. That is, the current study sought to manipulate children's underlying motivation by changing the purpose of the broader social situation – from learning through imitation to using what they learned about the box (in the canonical imitation situation) to accomplish their current goal of helping the adult. Therefore, it was hypothesized that over-imitation depends on the broader social situation such that children would be less likely to over-imitate in the helping situation relative to the canonical imitation situation.

5.2.4.4 Hypothesis 4

There was also a more subtle motivational manipulation that attempted to change the focus from within the imitation situation itself. There is some evidence that changes in the verbal framing of a demonstration can have consequences for children's understanding of its purpose (Kiraly, 2009; Herrmann, Legare, Harris, & Whitehouse, 2013). Therefore, by reframing the verbal instructions during the demonstration phase, the current study sought to highlight the *problem-solving* goal of opening the box in one condition while highlighting the goal of *playing* in the other condition. It was hypothesized that verbal framing would alter children's imitation behavior such that irrelevant actions would be copied more in the play condition and relevant actions would be copied more in the problem-solving condition.

In sum, the primary purpose of the current study was to explore the implications of characterizing imitation activity as a consequence of both the cognitive and motivational aspects involved with resolving uncertainty (via social learning) in different types of situations. Doing so involves trying to answer two intimately related issues in the literature: first, how to reconcile the apparent conflict between selective imitation and automatic over-imitation? Second, how to account for the underlying reason why children over-imitate?

5.2.5 *Methods*

5.2.5.1 Participants

Seventy-four, predominantly white, middle class children (47 male, 27 female) who were 3–5 years old (3-year-olds, $N = 27$, $M = 37$ months, SD = 1; 4-year-olds, $N = 21$, $M = 49$ months, $SD = 1$; 5-year-olds, $N = 20$, $M = 61$ months, $SD = 1$) participated in the study. Children were recruited through published birth announcements and from the university daycare. Of the seventy-four children, six children were excluded from coding and analyses (2 for being uncooperative and 4 for not wanting to take a turn opening the boxes).

5.2.5.2 Apparatus

Materials included some stuffed animals, two punch buttons, and two wooden boxes (30 cm × 30 cm × 20 cm) differing only in color. Each box had a drawer with a knob/handle on the front to open it. The drawer could be locked by a pin in the top of the box with a non-functional rotating lever attached behind the pin and rotated 180 degrees. Finally, a screen facade placed on top of the box and over the focal point of the lever also hid a surreptitious locking device on the side of the box.

5.2.5.3 Design

The design can be summarized as follows: 3 (age group: 3-, 4-, & 5-year-olds; between subjects) ×2 (verbal framing: play vs. problem-solving; between subjects) ×5 (demonstration: step1, step2, step3, step4, step5; within-subjects) ×2 (situation type: imitation vs. helping; within-subjects).

5.2.5.4 Procedure

The procedure involved a variation on the active helping paradigm (Buttelmann, Carpenter & Tomasello, 2009). While this procedure was not originally used to study imitation it was ideally suited for the current purposes in that it involved a *demonstration* phase on how to open two unfamiliar boxes followed by a *cooperation* phase where children made use of their new found knowledge in order to help the adult.

After a warm-up session with two experimenters (E1 & E2), the demonstration phase began and the child was seated on a cushion that was equidistant (approximately 2 m) from the two boxes that were themselves about 1 m apart. The two punch-buttons were placed approximately 30 cm away from the inside edge of the boxes, the drawers were open, and the locking pin for each box was on the floor directly in front of them (approximately 10 cm). E1 was seated next to the child while E2 was seated between the two boxes facing the child and E1. After exploring the boxes briefly, E2 left the room in search of their favorite toy while E1 demonstrated to the child how the boxes could be opened with two response trials for each box.

Children were randomly assigned to one of two conditions that manipulated the verbal framing of the demonstration. Specifically, the experimenter introduced the task either as an opportunity to learn how to *play* with the boxes "Let me show you how to play with the boxes" (play condition) or as an opportunity to learn how to *open* the boxes "Let me show you how to open the boxes" (problem-solving condition). E1 provided two demonstrations on the first box (counterbalanced) before the child was given the first of their two turns on that box. E1 provided one additional demonstration on the second box before the child was given the first of their two turns on that box. No reinforcement was given while children took their turns with the boxes. Finally, between all turns, E1 asked children to return to the pillow, to cover their eyes, and obstructed their view in order to guarantee that children did not see how the boxes were reset.

A 5-step procedure that was identical for both boxes was demonstrated to children: Two of these steps were causally relevant and three were causally irrelevant. The two relevant steps included step 3: pulling a "locking" pin out of a hole; and step 5: pulling on a knob/handle to open the drawer. The three irrelevant steps were designed to be differentially plausible. Of these, step 2 was intended to be the most implausible and involved rotating a lever on the top of the box from right to left 180°. Step 1 involved a punch button that was pushed once and step 3 involved tapping the "locking" pin on the edge of the box three times. Tapping on the box was thought to be less plausible than pushing the button but previous research has suggested that physical contact may mediate functional inferences (Lyons et al., 2007).

After the demonstration phase with E1, E2 returned to the room with their favorite toy and engaged the child in play. After 90 s, E2 placed the toy in one of the boxes (counterbalanced) and left the room. Next, E1 invited the child to play a trick on E2 by moving the toy to the other box and ensured that both boxes were reset (locked). At this point, the *cooperation* phase began. The transition in phases from demonstration (i.e., imitation situation) to cooperation (i.e., helping situation) constituted the within-subjects manipulation of motivation. Upon re-entering the room, E2 attempted to open the now empty box where they had previously left their toy. Unable to open the box, E2 sat in a position centered between the two boxes, faced the child and looked "disappointed", "puzzled", and "resigned" (Buttelmann et al., 2009). Next, the response period began in which the child was able to help E2 open either of the boxes; however, the boxes were surreptitiously locked and would only open if the child performed all five target-actions in the correct sequence. If children did not spontaneously try to help E2, then they were prompted to do so but were not given any cues about how to open the box (e.g., E1: "Go on, you can help him/her" and/or E2: "Can you help me" and/or E1: "We can go together and help him/her").

The procedure can be summarized as follows: During the demonstration phase of the experiment (i.e., the imitation situation), all children were given four demonstration and response trials for how to open the two identical boxes (evenly and randomly split between play and problem-solving conditions). Following this was the cooperation phase (i.e., the helping situation) in which children were given an opportunity to use what they learned about opening the boxes to help an adult confederate. However, the boxes were surreptitiously locked and would only open if all five target-actions were performed.

5.2.5.5 Coding and Reliability

All of the analyses concerning social-learning aspects of the study were based on the *first five actions* that children performed on the boxes averaged across the four training trials (2 for each box); however, this did not include consecutive repeats (e.g., steps 1,1,2,3 would be coded as 1,2,3 but 1,2,1 would be coded as 1,2,1). The resulting target action (TA) score was based on how many of the 5 target actions that children performed within the *first five actions* that they performed (max = 5, 1 for each of the distinct target actions, e.g., 1,2,1 would have two TAs whereas 1,2,3 would have three TAs). For hypothesis 2, concerning the influence of children's prior artifact knowledge, a total TA score was calculated on the bases of the *total* number of times that children produced any of the five distinct target actions throughout the demonstration phase; again, consecutive repeats were not included. There was no maximum because children were not forced to stop after their first five actions.

All of the data was recoded by the collective efforts of three observers who independently viewed the videos and were unaware of the specific hypotheses of the experiment. There was 91% agreement between the original and recoded data. Discrepancies were settled by reviewing the video recordings. It was typically clear that one of the coders had made a mistake rather than there being any disagreement on the status of an ambiguous action. For those few cases where the action was ambiguous, a third coder settled the disagreement.

5.2.6 *Results*

5.2.6.1 Hypothesis 1: Did Over-imitation Increase with Age?

In order to test hypothesis 1, a multivariate analysis of variance (MANOVA) was conducted to assess whether there were differences in the number of relevant and irrelevant target actions performed across age groups (3-, 4-, & 5-year-olds) and verbal framing conditions (play vs. problem-solving). Results indicated that imitation of relevant actions did not differ across the three age groups whereas imitation of irrelevant actions increased with age ($F(2, 62) = 10.25, p < .001, M_3 = 1.69, SD = .79; M_4 = 2.35, SD = .51; M_5 = 2.48, SD = .51$). Post hoc analyses (LSD) revealed significant differences between 3- and 4-year-olds ($p < .001$) and 3- and 5-year-olds ($p < .001$), but not between the two older groups (see Fig. 5.1).

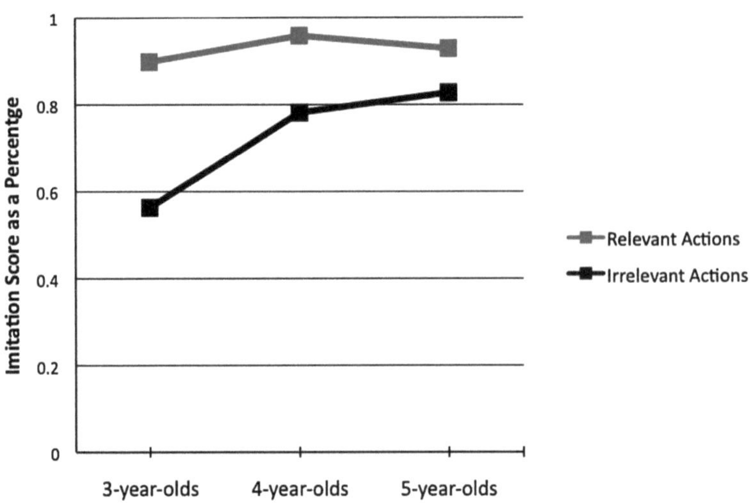

Fig. 5.1 Imitation scores for relevant and irrelevant target actions for 3-, 4-, and 5-year-olds. Children's imitation score is given in a percentage for clarity of comparison because there were only two relevant actions possible while there were three irrelevant actions possible. Post hoc analyses indicated that for irrelevant actions only, 4- and 5-year-old children differed significantly from 3-year-olds ($p < .001$ for both) but not from each other

5.2.6.2 Hypothesis 2: Were Children Sensitive to the Difference Between Relevant and Irrelevant Actions?

To test hypothesis 2, a mixed model analysis of variance (ANOVA) was conducted to assess whether there were differences in the total number of times each of the five target actions was performed at any point in the demonstration phase with age (3-, 4-, & 5-year-olds) and verbal framing (play or problem-solving) as between subjects variables. Although the assumptions of sphericity were violated, Greenhouse-Geisser corrections did not alter the statistical decision for any of the effects. The overall analysis indicated a significant effect of step ($F(3.22, 200.45) = 17.90, p < .001$). Main effects for age and for verbal framing were marginally significant ($F_{Age}(2, 62) = 2.54, p = .087, \eta^2 = .076; F_{Framing}(1, 62) = 3.25, p = .076, \eta^2 = .050$). Specifically, 3-year-olds copied fewer overall target actions than 4- and 5-year-olds and children in the problem-solving condition copied more overall target actions than those in the play condition.

None of the possible interactions approached significance. In particular, there was not a significant interaction between step and age ($F(6.47, 200.45) = 1.50, p = n.s.$) indicating that the relative differences between overall imitation of the five steps did not differ across the three age groups. Finally, follow-up analyses indicated that children preferentially copied those actions that were more likely to be relevant for actually opening the box. That is, all age groups tended to imitate the two relevant actions most followed by the irrelevant actions (see Table 5.1). Within-subject contrasts confirmed that children differentiated between the relevant and irrelevant target actions. Specifically, step 5 was significantly different from steps l, 2, and 4 ($p < .001$) but not step 3.

5.2.6.3 Hypothesis 3: Does Over-imitation Depend on the Broader Type of Situation?

This hypothesis was being tested against an alternative hypothesis that children over-imitate (imitate irrelevant actions) because they have incorporated those actions into their "causal belief structure" for how to open the box independently of the type of situation (Lyons et al., 2007, 2011). To test these alternative hypotheses, children's imitation activity on their last trial of the demonstration phase (imitation

Table 5.1 Average number of times that children produced each of the five distinct target actions at any point throughout the imitation phase

	M	SD	Relevant Action
Step 1 (button press)	3.29	1.37	
Step 2 (lever turn)	2.82	1.67	
Step 3 (pin pull)	4.04	1.27	*
Step 4 (tap 3 times)	2.49	1.73	
Step 5 (pull knob)	4.13	1.64	*

Note. Step 5 differed significantly from steps l, 2 & 4 ($p < .001$) but not step 3. The * indicates that the step was a relevant action

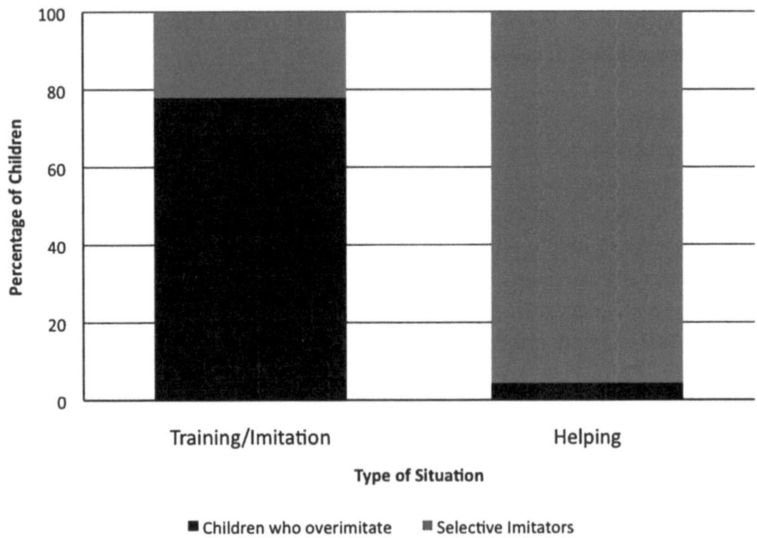

Fig. 5.2 Percentage of children who over-imitated in the imitation situation versus the helping situation of the experiment ($\chi^2 = 72.43, p < .001$)

situation) was compared to their first trial in the cooperation phase (helping situation). In both cases their actions were only coded until they attempted to open the box by pulling on the knob (i.e., step 5). Four children did not want to respond during the helping phase of the experiment and were not included in the analysis of that phase. Children were categorized as over-imitators if they performed two or more of the irrelevant actions. A Chi-Square analysis indicated that significantly more children over-imitated in the imitation situation ($N = 53/68 = 78\%$) than in the helping situation ($N = 3/64 = 5\%$; $\chi^2 = 72.43, p < .001$, see Fig. 5.2).

Follow up analyses indicated that 3-, 4-, and 5-year-olds all differed significantly in their over-imitation behavior across the two situations ($p < .001$). That is, all three age groups demonstrated a significant decline in over-imitation from the imitation to the helping situation (see Fig. 5.3). However, 3-year-olds differed significantly from 4- & 5-year-olds in terms of the proportion of children who over-imitated in the imitation situation versus the helping situation ($\chi^2 = 13.17, p = .001$). While just over half (56%) of 3-year-olds over-imitated in the imitation situation, 90% of 4-year-olds and 95% of 5-year-olds over-imitated in the imitation situation (see Fig. 5.3).

Although children aged 3–5 and younger should not have a problem remembering a 5-step sequence (Bauer & Fivush, 1992; McGuigan & Whiten, 2009), we ensured that children who did not over-imitate in the helping situation could demonstrate that they had not forgotten the irrelevant steps. After failing to open the box using the two causally relevant actions (it was surreptitiously locked), 75% of children made spontaneous use of the irrelevant actions.

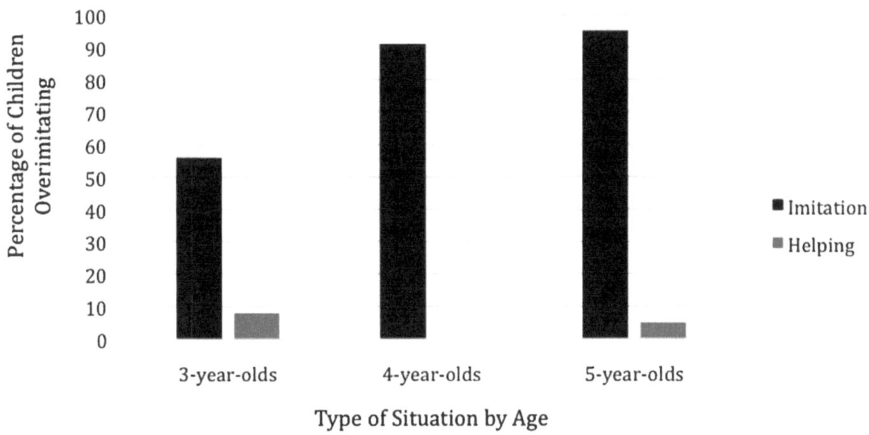

Fig. 5.3 Percentage of children who over-imitated in the imitation situation versus the helping situation divided by age group

5.2.6.4 Hypothesis 4: Does Verbal Framing Effect Imitation Response?

The analysis for the age hypothesis (hypothesis 1) assessed whether there were differences in the number of relevant and irrelevant target actions performed across age groups and verbal framing conditions (play vs. problem-solving). Results for the verbal framing variable revealed a marginal main effect for the imitation of *relevant* ($F(1, 62) = 3.87, p = .054$) but not *irrelevant* ($F(1, 62) = .57, p = n.s.$) target actions. Specifically, children in the problem-solving condition imitated the relevant actions more often ($M = 1.91, SD = .16$) than did children in the play condition ($M = 1.78, SD = .34$).

5.2.6.5 Summary

First, over-imitation increased with age. Second, children differentially imitated the relevant versus irrelevant actions. Third, what actions children copied changed with the type of situation such that they over-imitated significantly less in the helping situation. Fourth, verbal framing had a moderately significant effect for relevant but not irrelevant actions.

5.2.7 *Discussion*

The purpose of the current study was to explore two intimately related issues in the literature: first, how to reconcile the apparent conflict between selective imitation and automatic over-imitation? Second, how to account for the underlying reason why children over-imitate?

Three- to 5-year-old children come to an experimental imitation situation with a rich developmental history of learning and experience interacting with both physical objects and social persons. Social interactions inherently involve a broader type of situation and when these situations involve novel aspects there will be uncertainty about how to proceed; however, such uncertainty will not be absolute. That is, in a social situation involving novel aspects, how children proceed will be heuristically guided by their prior knowledge and learning about objects and people in different types of situations. Natural Pedagogy has aptly demonstrated that canonical imitation experiments have a number of social-communicative markers that orient children's understanding about the type of situation that they are involved with (Csibra & Gergely, 2009). This orienting helps them to interpret the purpose of the situation and facilitates how to act "successfully" when given their turn. Accordingly, in a canonical imitation situation involving a relatively novel artifact, selecting what to copy from an adult demonstration will involve three interrelated aspects: social-cognitive, cognitive and motivational.

5.2.7.1 Beyond Artifacts and Agents: Social-Cognitive and Motivational Aspects

A major result from the current study (i.e., hypothesis 3) was the large decrease in the number of children who over-imitated in the helping situation relative to the imitation situation. Nearly 80% of children in the demonstration phase over-imitated, while this number dropped to less than 5% in the cooperation phase. This finding is in stark contrast to Lyons and colleagues' (2007, 2011) efforts to demonstrate that over-imitation involves a stable distortion to the child's causal-belief structure about how the artifact operates.

Cognitive-Learning Perspective

While the current results are consistent with the conclusion that children are learning *something* about the box during the imitation situation, we disagree that children are learning about the "causal structure" of the box.[5] Instead, it is a type of interactive or user knowledge similar to the type of learning typical of most adults for how to operate a car (i.e., only the mechanically educated know that the gas pedal controls an air valve, that gears do not move such that they could grind, and that engines cannot flood by pushing the gas pedal too much). From the current perspective, children are learning how to successfully interact in a social situation. Consequently, learning with respect to the box per se is tied to what children are learning about the social interaction that, as a whole, defines the situation.

[5]See Want and Harris (2001) for an example of a research design that tests for a causal learning interpretation for how to operate a novel artifact.

In general, the relevant interactive affordances of an artifact/object are going to change depending on the type of situation that is involved (e.g., consider the different affordances between the legendary Wayne Gretzky's hockey stick in the hockey hall of fame and the "same" one sitting on the bench at a local ice-skating rink). One methodological consequence of excluding this context-*dependent* nature for object affordances is that researchers have been overly focused on the "objective/causal" structure of the box independently of the social situation. However, what is relevant about an object depends on one's purposes and these will vary in different types of situations. Therefore, future research may benefit from incorporating the child's purpose/motivation for engaging with an object in a way that is tied to their understanding of the broader type of situation.

Social-Motivational Perspective

The results from hypothesis 3 provided evidence that motivational aspects are an essential variable for understanding why children over-imitate. However, fully recognizing the relevance of motivational aspects for over-imitation requires incorporating a broader appreciation for the heterogeneity of social situations in general. Knowledge about different types of social situations (and any corresponding motivations) must be learned and that forces a theoretical focus on children's developmental history of learning how to competently interact with other people in different types of situations.

One methodological consequence of excluding the broader social situation for the social-motivational perspective is that it has been overly focused on whether social agents are present or absent for some aspect of the experiment. While agents are certainly part of what makes a situation social in general, it is the commonly held understanding of how to engage with each other that makes it a social reality (Bickhard, 2008). Accordingly, this means that the presence of an agent may be most relevant for the learning of the social reality but is not essential for manifesting such knowledge once learned (e.g., other people need not be around to manifest the rituals of religion). Therefore, future research may benefit from focusing on the nature of these commonly held understandings and how they develop into types of social situations.

In sum, findings from the current study provided converging evidence that over-imitation in preschoolers is less about the causal structure of the box and more about the social-cognitive structure of the situation as a whole (Kenward, Karlsson, & Persson, 2011; Kiraly et al., 2013). This social-cognitive structure helps children to understand the purpose of the situation (motivational aspects) as well to define the meaning of the object in the situation (cognitive aspects). To put it simply, children are learning that "this is what you do to the artifact to participate in the social reality" rather than that "this is the necessary causal structure for how to open the box".

5.2.7.2 Re-thinking Novelty: Cognitive Aspects

The results from hypotheses 1 and 2 suggest that children's existing knowledge about generic objects influences their presumptions about how a relatively novel box might operate? Whatever children (and adults) are able to learn through imitation, it must be within the context of what they already know about the world. That is, all imitation learning operates within the child's zone of proximal development. Imitation learning enables children to extend the boundaries of what they already know but does not enable learning something that is entirely novel. This may be both because of needing to know where to deploy attention when perceiving the demonstration but also because of needing to know enough about the domain to recreate the target actions. Learning to play sports by imitation highlights the limitations of both perceiving what actions the demonstrator provided and (unfortunately for aspiring sports stars) how to recreate them.

The results from hypothesis 2 (see Table 5.1) indicated differences in the total number of times that each of the 5 steps was reproduced at *any* point throughout the training phase. The step (step 4–tapping) that was least likely to be relevant for opening the box was produced the least followed by the other two, more likely, but also irrelevant steps (steps 1–button press & 2–lever rotation), with the two relevant steps (steps 3–pin pull & 5–pull knob) being imitated the most. We interpret this differential behavior as a consequence of children's zone of proximal development with respect to their knowledge about container like objects. That is, children with more robust object knowledge can better copy the less plausible target actions from the demonstration. In contrast, those children with less robust object knowledge may be more focused on learning about the box as physical object. This interpretation accords well with the developmental data from hypothesis 1 in which 3-years can be presumed to have less robust object knowledge than the 4- and 5-year-olds.

5.2.7.3 Transcending Fidelity: Cognitive and Motivational Aspects

The results of hypothesis 1 indicated that *relevant* target actions were imitated uniformly across the three age groups while there was a developmental increase in children's imitation of *irrelevant* target actions between 3- and 4-years of age (Fig. 5.1). Consider that age differences are acting as a surrogate for task difficulty and subsequent degree of uncertainty. Harnick (1978) demonstrated that imitation behavior in toddlers is influenced by task difficulty such that fidelity was greatest when the task was of moderate difficulty (i.e., not too difficult or too easy). Such a dynamic suggests that there are relevant motivational-learning processes involved with imitation learning (Kaplan & Oudeyer, 2007). In the current task there is some amount of uncertainty for learning about both the object and the social reality involving the interaction with the adult. Therefore, the relative lack of over-imitation by the 3-year-olds suggests that they may have been motivated to learn more about the box than the broader social reality (about half did not over-imitate). In contrast, learning about the box may have been sufficiently trivial for the 4- and 5-year-olds such that they were motivated to learn about the more complicated social reality involving the box as a cultural artifact.

The suggested interpretation of these results does not propose an absolute claim about children of these ages engaging in a *type* of imitation (i.e., selective, over-, goal-, means-, social, cognitive, etc., Dickerson, Gerhardstein, Zack & Barr, 2012; Over & Carpenter, 2012). Instead, there is a task dependent dynamic that plays out for all imitation learning in which children imitate what they understand to be relevant about the demonstration and their determination about what is relevant depends on social-cognitive, cognitive, and motivational processes.

5.2.7.4 Verbal Framing and Forgetting

The results for hypothesis 4 indicated that the effect of verbal framing on children's imitation behavior was modest. Children in the *problem-solving* condition imitated with higher fidelity than those in the *play* condition; however, this difference only approached significance with respect to relevant target actions. Further, there was not the expected increase in imitation of irrelevant target actions for the play condition. Given some recent evidence for an effect of verbal framing on imitation fidelity (Herrmann et al., 2013), we believe that the subtlety of the effects in the current study are due to the weakness of the manipulation itself. In the current study: the transition from the warm-up period involved referring to the imitation task as the "boxes game" for both conditions; the task was presented with a playful and game-like structure; and, the objects involved were relatively unfamiliar (cf., Kiraly 2009; Herrmann et al. 2013). As for the manipulation, the only difference between conditions was whether the experimenter used the word "play" versus "open" when introducing the task (e.g., "Let me show you how to [play with/open] the boxes. I'm going to take two turns and then you'll get two turns"). Therefore, aspects common to both conditions contributed to the situation being about *playing* games and suggests that a stronger manipulation may find an effect.

5.3 Conclusion

The current study was meant to address two intimately related issues in the literature: First, how to reconcile the apparent conflict between selective imitation and automatic over-imitation? Second, how to account for why children over-imitate? The analysis did not answer the first question directly but rather argued for an alternative characterization that highlights the selective nature of all forms of imitation activity. Consequently, the current study proceeded to explore the psychological basis for selectivity in a canonical over-imitation setting. In particular, how cognitive and motivational aspects were related to the broader type of social situation and how they collectively guided children's imitation activity. The results support our argument that children's over- and selective imitation depends on the specifics of the social situation. Specifically that is, we conclude that children over-imitate because they are motivated to learn how to competently interact with people and objects in different types of social situations. That is, children over-imitate because it is the nature of learning about social realities.

What is special about imitation learning? Imitation is widely assumed to be a particularly important form of social learning because of its contribution to human culture more broadly. However, understanding of *cultural* artifacts does not reduce to the causally necessary ways of using objects or even to their instrumental functioning. Instead, cultural artifacts are saturated with aesthetics and their functioning is tied to our social interests as much as to our "rational" needs. Cultural artifacts are constrained by their physical embodiment but attain social meaning through social agents participating in social realities. Likewise, social ontology does not reduce to the presence or absence of social objects that are understood as more complex physical objects. Instead, social ontology is emergent from interactive systems grounded in the world. Social realities are fundamentally constrained by coordination and although they will involve arbitrary aspects (e.g., it does not matter what side of the road we drive on) that does not mean that there are no normative constraints for understanding what one ought to do to participate in cultural activities (i.e., rationality and necessity do not exhaust the normative ground for imitation learning).

The current action-based perspective offers a robust theoretical framework for exploring some of the richness of learning to participate in human culture. This framework draws on a model of the basic content of culture in terms of social ontology and provides a genuinely developmental perspective on those learning to participate in such activities. Standard frameworks in developmental psychology do not motivate consideration of how learning processes themselves must undergo learning and development. To the extent learning to learn is essential for the development of human forms of activity, it may be a productive avenue for roboticists to explore in order to realize the full potential of artificial systems. Further, a learning to learn framework highlights the need to consider the development of emergent forms of motivation, including social motivations, and helps explain why motivations for different forms of learning change with development (e.g., communicative imitation tends to decline around age 4 with the increased availability of learning through language, Nadel, 2006). Finally, standard frameworks define social ontology in terms of the ability to attribute mental-state contents to self and other (i.e., mindreading, theory of mind). However, the emergent origins of such mental-state contents (as well as the attribution process) remain a mystery (Carpendale & Lewis, 2006). This does not mean that a mindreading framework toward social ontology cannot be implemented in robotics models (Dominey & Warneken, 2010), but rather, that there are serious questions about whether that approach can adequately capture the nature, origins, and development of social robots (Thompson, Sameen, Maximilian, & Racine, 2013).

In sum, the current empirical work and theoretical integration constitute a first step for a promising new avenue of research in the domain of social meta-learning that roboticists may well need to consider if they are to build social agents capable of participation in human socio-cultural activities.

Acknowledgements We would like to thank Mark Bickhard and Ulrich Müller for useful comments throughout writing the manuscript. We would also like to thank Andrea Lucas, Mitchel Harvey, and Bryan Hughes for help with data collection.

References

Allen, J. W. P., & Bickhard, M. H. (2011a). Emergent constructivism. *Child Development Perspectives, 5*(3), 164–165.

Allen, J. W. P., & Bickhard, M. H. (2011b). You can't get there from here: Foundationalism and development. *Behavioral and Brain Sciences, 34*, 124–125.

Allen, J. W. P., & Bickhard, M. H. (2013a). Beyond principles and programs: An action framework for modeling development. *Human Development, 56*, 171–177.

Allen, J. W. P., & Bickhard, M. H. (2013b). Stepping off the pendulum: Why only an action-based approach can transcend the nativist-empiricist debate. *Cognitive Development, 28*, 96–133.

Anisfeld, M. (2005). No compelling evidence to dispute Piaget's timetable of the development of representational imitation in infancy. In S. Hurley & N. Chater (Eds.), *Perspectives on imitation: From neuroscience to social science* (Vol. 2, pp. 107–131). Cambridge: MIT Press.

Bauer, P. J., & Fivush, R. (1992). Constructing event representations: Building on a foundation of variation and enabling relations. *Cognitive Development, 7*, 381–401.

Bickhard, M. H. (1980). *Cognition, convention, and communication*. New York: Praeger Publishers.

Bickhard, M. H. (2005). Functional scaffolding and self-scaffolding. *New Ideas in Psychology, 23*, 166–173.

Bickhard, M. H. (2008). Social ontology as convention. *Topoi, 27*, 139–149.

Bickhard, M. H. (2009). The interactivist model. *Synthese, 166*, 547–591.

Bickhard, M. H. (in preparation). *The whole person: Toward a naturalism of persons—Contributions to an ontological psychology*. (In preparation)

Bickhard, M. H., & Terveen, L. (1995). *Foundational issues in artificial intelligence and cognitive science: Impasse and solution*. Amsterdam/New York: Elsevier Scientific.

Buttelmann, D., Carpenter, M., & Tomasello, M. (2009). Eighteen-month-old-infants show false belief understanding in an active helping paradigm. *Cognition, 11*, 337–342.

Call, J., & Carpenter, M. (2002). Three sources of information in social learning. In K. Dautenhahn & C. L. Nehaniv (Eds.), *Imitation in animals and artifacts* (pp. 211–228). Cambridge: MIT Press.

Call, J., & Carpenter, M. (2009). Monkeys like mimics. *Science, 325*, 824–825.

Campbell, R. L., & Bickhard, M. H. (1992). Type of constraints on development: An interactivist approach. *Developmental Review, 12*, 311–338.

Carpendale, J. I. M., & Lewis, C. (2006). *How children develop social understanding: Understanding children's worlds*. Oxford: Blackwell.

Corriveau, K., & Harris, P. L. (2009). Choosing your informant: Weighing familiarity and recent accuracy. *Developmental Science, 12*, 426–437.

Csibra, G., & Gergely, G. (2009). Natural pedagogy. *Trends in Cognitive Science, 13*, 148–153.

Dickerson, K., Gerhardstein, P., Zack, E., & Barr, R. (2012). Age-related changes in learning across early childhood: A new imitation task. *Developmental Psychobiology, 55*, 719–732.

Dominey, P. F., & Warneken, F. (2010). The basis of shared intentions in human and robot cognition. *New Ideas in Psychology, 29*, 260–274.

Gergely, G. (2003). The development of teleological versus mentalizing observational learning strategies in infancy. *Bulletin of the Menninger Clinic, 67*, 113–131.
Gergely, G., Bekkering, H., & Kiraly, I. (2002). Rational imitation in preverbal infants: Babies may opt for a simpler way to turn on a light after watching an adult do it. *Nature, 415*, 755.
Gergely, G., & Csibra, G. (2005). The social construction of the cultural mind: Imitative learning as a mechanism of human pedagogy. *Interaction Studies, 6*, 463–481.
Harnick, F. S. (1978). The relationship between ability level and task difficulty in producing imitation in infants. *Child Development, 49*, 209–212.
Harris, P. L., Corriveau, K., Pasquini, E. S., Koenig, M., Fusaro, M., & Clément, F. (2012). Credulity and the development of selective trust in early childhood. In M. J. Beran, J. Brandl, J. Perner, & J. Proust (Eds.), *Foundations of metacognition* (pp. 193–210). Oxford, England: Oxford University Press.
Herrmann, P. A., Legare, C. H., Harris, P. L., & Whitehouse, H. (2013). Stick to the script: The effect of witnessing multiple actors on children's imitation. *Cognition, 129*, 536–543.
Heyes, C. (2005). Imitation by association. In S. Hurley & N. Chater (Eds.), *Perspectives on imitation: From neuroscience to social science* (Vol. 1, pp. 157–176). Cambridge: MIT Press.
Horner, V., & Whiten, A. (2005). Causal knowledge and imitation/emulation switching in chimpanzees (pan troglodytes) and children (homo sapiens). *Animal Cognition, 8*, 164–181.
Huang, C. T., Heyes, C., & Charman, T. (2006). Preschoolers' behavioral reenactment of "failed attempts": The roles of intention-reading, emulation and mimicry. *Cognitive Development, 21*, 36–45.
Jones, S. S. (2007). Imitation in infancy: The development of mimicry. *Psychological Science, 18*, 593–599.
Kaplan, F., & Oudeyer, P. (2007). The progress drive hypothesis: An interpretation of early imitation. In K. Dautenhahn & C. Nehaniv (Eds.), *Models and mechanisms of imitation and social learning: Behavioral, social and communication dimensions* (pp. 361–377). Cambridge: Cambridge University Press.
Kenward, B., Karlsson, M., & Persson, J. (2011). Over-imitation is better explained by norm learning than by distorted causal learning. *Proceedings of the Royal Society B, 278*, 1239–1246.
Kiraly, I. (2009). The effect of the model's presence and of negative evidence on infants' selective imitation. *Journal of Experimental Child Psychology, 102*, 14–25.
Kiraly, I., Csibra, G., & Gergely, G. (2013). Beyond rational imitation: Learning arbitrary means actions from communicative demonstrations. *Journal of Experimental Child Psychology, 116*, 471–486.
Lyons, D. E. (2009). The rational continuum of human imitation. In J. A. Pineda (Ed.), *Mirror neuron systems: The role of mirroring processes in social cognition* (pp. 77–103). New York: Humana Press.
Lyons, D. E., Damrosch, D. H., Lin, J. K., Deanna, M. M., & Keil, F. C. (2011). The scope and limits of over-imitation in the transmission of artifact culture. *Philosophical Transactions of the Royal Society B, 366*, 1158–1167.
Lyons, D. E., Young, A. G., & Keil, F. C. (2007). The hidden structure of overimitation. *Proceedings of the National Academy of Sciences, 104*(50), 19751–19756.
McCarthy, J., & Hayes, P. (1969). Some philosophical problems from the standpoint of artificial intelligence. In B. Meltzer & D. Michie (Eds.), *Machine intelligence 4* (pp. 463–502). New York: American Elsevier.
McGuigan, N., Makinson, J., & Whiten, A. (2011). From overimitation to super-copying: Adults imitate causally irrelevant aspects of tool use with higher fidelity than young children. *British Journal of Psychology, 102*, 1–18.
McGuigan, N., & Whiten, A. (2009). Emulation and "overemulation" in the social learning of causally opaque versus causally transparent tool use by 23- and 30-month-olds. *Journal of Experimental Child Psychology, 104*, 367–381.
McGuigan, N., Whiten, A., Flynn, E., & Horner, V. (2007). Imitation of causally opaque versus causally transparent tool use by 3- and 5-year-old children. *Cognitive Development, 22*, 353–364.

Meltzoff, A. N. (1995). Understanding the intentions of others: Re-enactment of intended acts by 18-month-old children. *Developmental Psychology, 31*, 838–850.

Nadel, J. (2006). Does imitation matter to children with autism? In S. J. Rogers & J. H. G. Williams (Eds.), *Imitation and the social mind: Autism and typical development* (pp. 118–137). New York: Guilford Press.

Nehaniv, C. L., & Dautenhahn, K. (2002). The correspondence problem. In K. Dautenhahn & C. L. Nehaniv (Eds.), *Imitation in animals and artifacts* (pp. 41–62). Cambridge: MIT Press.

Nielsen, M. (2006). Copying actions and copying outcomes: Social learning through the second year. *Developmental Psychology, 42*, 555–565.

Nielsen, M., & Blank, C. (2011). Imitation in young children: When who gets copied is more important that what gets copied. *Developmental Psychology, 47*, 1050–1053.

Nielsen, M., Simcock, G., & Jenkins, L. (2008). The effect of social engagement on 24-month-olds' imitation from live and televised models. *Developmental Science, 11*, 722–731.

Over, H., & Carpenter, M. (2012). Putting the social into social learning: Explaining both selectivity and fidelity in children's copying behavior. *Journal of Comparative Psychology, 126*, 182–192.

Paulus, M., Hunnius, S., Vissers, M., & Bekkering, H. (2011). Imitation in infancy: Rational or motor resonance? *Child Development, 82*, 1047–1057.

Piaget, J. (1954). *The construction of reality in the child*. New York: Basic.

Schulz, L. E., Hooppell, C., & Jenkins, A. C. (2008). Judicious imitation: Children differentially imitate deterministically and probabilistically effective actions. *Child Development, 79*, 395–410.

Southgate, V., Chevallier, C., & Csibra, G. (2009). Sensitivity to communicative relevance tells young children what to imitate. *Developmental Science, 12*, 1013–1019.

Thompson, J. J., Sameen, N., Maximilian, B., & Racine, T. P. (2013). Agnosticism gone awry: Why developmental robotics must commit to an understanding of embodiment and shared intentionality. *New Ideas in Psychology, 31*, 184–193.

Tomasello, M. (1996). Do apes ape? In C. M. Heyes & G. Galef (Eds.), *Social learning in animals: The roots of culture* (pp. 319–346). San Diego, CA: Academic Press.

Uzgiris, I. C. (1981). Two functions of imitation during infancy. *International Journal of Behavioral Development, 4*, 1–12.

Want, S. C., & Harris, P. L. (2001). Learning from other people's mistakes: Causal understanding in learning to use a tool. *Child Development, 72*, 431–443.

Whiten, A. (2006). The dissection of imitation and its "cognitive kin" in comparative and developmental psychology. In S. J. Rogers & J. H. G. Williams (Eds.), *Imitation and the social mind: Autism and typical development* (pp. 277–309). New York: Guilford Press.

Whiten, A., McGuigan, N., Marshall-Pescini, S., & Hopper, L. M. (2009). Emulation, imitation, over-imitation and the scope of culture for child and chimpanzee. *Philosophical Transactions of the Royal Society B, 364*, 2417–2428.

Williamson, R. A., & Markman, E. M. (2006). Precision of imitation as a function of preschoolers' understanding of the goal of the demonstration. *Developmental Psychology, 42*, 723–731.

Zentall, T. R. (2006). Imitation: Definitions, evidence, and mechanisms. *Animal Cognition, 9*, 335 353.

Chapter 6
Mindshaping and Robotics

Víctor Fernández Castro

Abstract Social robotics attempts to build robots able to interact with humans and other robots. Philosophical and scientific research in social cognition can provide social robotics research with models of social cognition to implement those models in mechanic agents. The aim of this paper is twofold: firstly, I present and defend a framework in social cognition known as mindshaping. According to it, human beings are biologically predisposed to learn and teach cultural and rational norms and complex cultural patterns of behavior that enhance social cognition. Secondly, I will highlight how this framework can open new research perspectives in the area of social robotics.

Keywords Mindreading • Mindshaping • Normativity • Situated robotics

6.1 Social Robotics and the Research in Social Cognition

Social robotics is a new area of research framed into the field of artificial intelligence. The primary objective of social robotics is developing robots able to interact with humans or with other robots. In spite of being a young field, social robotics has already proven to be a productive area with different branches such as education, gaming, companion or therapy (Breazeal, 2003). For instance, research in social robotics is leading to the development of robots which are used in therapy with autistic children (Cabibihan, Javed, Ang Jr, & Aljunied, 2013). Such robots are capable of interacting with the children to improve their social skills and emotional responses.

Diverse groups of paradigms converge in social robotics, such as artificial life, humanoids or mobile robots; and different fields of application, for instance, therapy, education, household or scientific research. This diversity makes the definition of social robotics elusive (Moral, Pardo, & Angulo, 2009) because we must pay attention to very different features of social interactions in order to

V. Fernández Castro (✉)
Dpto. Filosofía I, Edificio Psicología, Universidad de Granada, 18071 Granada, Spain
e-mail: vfernandezcastro@gmail.com

R. Hakli, J. Seibt (eds.), *Sociality and Normativity for Robots*, Studies in the Philosophy of Sociality 9, DOI 10.1007/978-3-319-53133-5_6

implement robots. In this sense, various scholars have emphasized the importance of very different aspects: agency, embodiment, emotion or communication. For example, Fong, Nourbakhsh, and Dautenhahn (2003) argue that social robots should be embodied in the same environment as human agents; while Breazeal (2003) highlights the importance of agency. Depending on those aspects, the development of social robots could be radically different. The task of defining social interactions in order to develop social robotics requires a combination of conceptual exercise with empirical research. This exemplifies the necessity of including conceptual clarification of human social interactions intending to enhance social robotics development.

In this line, philosophy of mind and cognitive sciences can provide different conceptual frameworks to tackle social interactions. Those disciplines are useful instruments to shed light on the mechanisms underlying social interaction. Thus, they can provide social robotics with different approaches and theoretical backgrounds in order to design robots. Moreover, this influence is bidirectional. Social robotics is a unique opportunity for philosophers to test and analyze the depth of their theories through their instantiations in mechanical agents. Following this line of reasoning, my aim is to present a new framework in social cognition that can contribute to the interdisciplinary relationship between social robotics and philosophy of mind and cognitive sciences.

6.2 Social Cognition: Mindreading or Mindshaping?

Among the unique cognitive capacities of human beings, the ability to navigate the social world is probably the most remarkable. Human beings expend the majority of their time engaged in social situations interacting with each other. They encounter other humans and animals with an unusual sensitivity, a sensitivity that substantially differs from the way they interact with physical objects. Furthermore, this unique expertise to confront social situations is translated into a high grade of virtuosity and complexity in cooperation, imitation, cultural learning and other social skills that are considerably rare in comparison to other species. During several decades, it has been widely agreed that the linchpin of this social proclivity is the faculty known as *mindreading*.

Mindreading is the capacity to understand other agents in mental terms. By ascribing a different array of mental states to others, we can understand, explain and predict overt behaviors in a way unavailable to other species. The orthodox view has centered social cognition research in the study of mindreading; those mentalizing skills are considered the linchpin of human social cognition (Baron-Cohen, 1999; Mithen, 1996; Dunbar, 2000, 2003). As a result, the primary focus of debate concerning mindreading has been the nature of the mechanism underlying the process. On the one hand, the theory-theory view claims that mental state attribution is produced by a kind of theorization based on a tacit theory of human mind (Carruthers, 1996; Nichols & Stich, 2003). On the other hand, simulationism claims that the process is carried out by a simulation mechanism based on introspections (Gordon, 1996; Goldman, 2006). In spite of the disagreement, both contenders assume that attributing mental states consist in postulating some non-observable inner states that cause behavior and enable its explanation and prediction.

Contrary to the orthodoxy, the perspective I canvass in this paper maintains that our primary socio-cognitive capacity is to form and acquire norms and patterns of behavior that regulate our social situations in order to make us more predictable (Andrews, 2015; Mameli, 2001; McGeer, 2007, 2015; Zawidzki, 2008, 2013). According to this view, humans are equipped with a set of *mindshaping*[1] mechanisms that facilitate the acquisition of a background of norms and routines that structure our social interactions. Anticipatory capacities derivate from the acquisition of those routines and norms, since they homogenize the behaviour of the population and make their behavior more predictable.

Understanding social cognition from one perspective or another has different implications for social interaction. In Sect. 6.4, I explore the implication of the mindshaping view for social robotics. Before that, I analyze both the mindreading and mindshaping views in some depth, and I present some arguments supporting the mindshaping view.

6.2.1 Mindreading

A common assumption in social cognition research is that the cornerstone of social abilities is the capacity of explaining and predicting behavior by attributing mental states. The rationale behind this view is that social circumstances require the agents to coordinate their actions in sophisticated ways that demand anticipatory and explanatory skills. Attributing the appropriate inner causes of the target's behavior allows us to make predictions about subsequent actions. Thus, mindreading facilitates other socio-cognitive capacities. For instance, mindreading aid to coordinate our actions to implement cooperative projects or enable complex communication by helping to track communicative intentions to access the meaning of a given utterance.

During several decades, social cognition research has been centered on the nature of the mechanisms underlying mindreading. The basic contenders are divided into two groups: the *theory-theories* and *simulationist* theories. According to the former, the psychological mechanism underlying mindreading is a tacit theory of the structure of human mind that helps us to infer mental states from behavior:

[1]There are several differences concerning the way those authors approximate those basic ideas. For instance, while McGeer (2007) dubs it the regulative view and emphasize the different ways we maintain the norms and routines, Zawidzki (2013) focuses on the developmental and evolutionary aspects of it. Although I appeal to several of those authors, during my exposition, I mostly elaborate from Zawidzki's ideas.

> [...] a specialised theory of mind module, which underlies the mind-reading abilities of normal subjects. This module is held to contain an implicit theory of the structure and functioning of human mind, which is accessed whenever a normal subject ascribes a mental state to another person, or seeks a mentalistic explanation of their behaviour. (Carruthers, 1996)

According to this view, mindreading is a question of theoretical reasoning. We infer others' mental states by using a theoretical apparatus implemented in our mind in order to make predictions of others' behavior.

On the other hand, according to the simulationist theories, the attribution of mental states is based on the capacity of simulating them. We use our motivational, emotional and practical-reasoning capacities in order to put ourselves in the shoes of the others:

> The recognition of one's own mental states is thought to be grounded in introspective access to these states, or at least in comparison of their qualitative features with a standard held in memory; and this is thought to require possession of the relevant mental state concepts. (Gordon, 1996)

We represent the mental states of others by simulating them in our mind. For example, we can predict what someone does in a particular situation by simulating what we would have done in the same situation. We can use our own cognitive resources in order to simulate others' mental states.

In spite of the differences, there are two basic assumptions that both simulationist theories and theory-theories widely accept:

The assumption of centrality: Mindreading is the linchpin of social cognition; therefore, we need the capacity of mindreading to perform any skill associated with the human socio-cognitive syndrome.

The representational assumption: The unique human capacity to explain others' behavior depends on the ability to represent others' mental states. In other words, we need an internal model or simulated representations of the structure of the mind for predicting courses of behaviors.

The assumption of centrality can be found explicitly maintained, for instance, in the work of Fodor (1987):

> What's relevant to whether common-sense psychology is worth defending is its dispensability in fact. And here the situation is absolutely clear. We have no idea of how to explain ourselves to ourselves except in a vocabulary which is saturated with belief/desire psychology. (Fodor, 1987, 9)

According to Fodor, there is no way we can explain each other without understanding the mental concepts. As he suggests the disappearance of mentalizing would be "the greatest intellectual catastrophe in the history of our species." (Fodor, 1987, xii). Nevertheless, the commitment to the assumption in the philosophy of mind and cognitive sciences is mostly tacit. For instance, like when socio-cognitive capacities as following gaze or joint attention are considered precursors of mindreading[2] (Baron-Cohen, 1994; Heal, 2005).

Likewise, the representational is widely established in several areas concerning the study of social cognition such as philosophy of mind or developmental psychology. For instance, Forguson and Gopnik (1988) express it when talking about the acquisition of social cognition in infants:

> The 4-year-olds have developed a representational model of the mind. This model construes the relation between the mind and external reality as mediated by mental representations: mental states with contents that have satisfaction conditions in the external world. Some of these states are satisfied (roughly: the world is as it is represented as being); some of them are not. The world is independent of our thought and experience; but it is represented in thought and experience. (Forguson & Gopnik, 1988, 236)

Humans need an inner representational model of the mind in order to interpret other creatures; they necessitate to comprehend that others' behavior are brought out by some representational states.

In order to understand how both assumptions play a role in the explanation of social interactions, consider the following examples. Firstly, imagine a scenario where two hunters are trying to capture a prey. Hunter_1 approaches the prey with a spear. Hunter_2 is also approaching the prey from the opposite side. Hunter_2 accelerates his pace. When Hunter_1 sees Hunter_2 approaching the prey from the opposite side, he expects the prey to respond by fleeing toward her position. Then, she waits crouched down to attack the prey when it flees toward her. Secondly, consider a scenario where you are a bartender in an Irish Pub. The Pub is not busy, only a few customers. Now that the Pub is almost empty, you wash the floor. Then, you see that one of the customers stands up from the seat and he approaches the bar while picking up his wallet. You expect him to pay, so you go straight to the cash register to give him the bill.

The way to explain those cases from the mindreading perspective is through postulating mental states conceived as representations. In the first example, Hunter_1 attributes to the Hunter_2 the intention of capturing the prey. Then, he attributes to the prey the state of fear that provokes its fleeing response, and then, he acts according to it. The second case is similar; you attribute to the costumer the intention to pay when you see him approaching the bar picking the wallet. Then you act in consequence. In both cases, by postulating inner representational states that bring out a particular course of behavior, the interpreter generates certain expectations that help him to coordinate his behavior with the target.

---

[2]In spite of the agreement, an increasing number of dissenters against centrality have emerged in the recent years (Gallagher, 2001; Hutto & Ratcliffe, 2007; Leudar & Costall, 2009). All these scholars share their refusal to the importance that the orthodoxy has assigned to mentalizing in the explanation of social cognition. This set of views is often called *interactionism* (Gallagher, 2004). According to it, social exchanges are usually facilitated for basic forms of socio-cognitive mechanisms including perceiving bodily movement as a goal-directed intentional movement, coordinating expressions and gestures or following gaze mechanisms.

6.2.2 Mindshaping

In contrast with the mindreading view, the mindshaping approximation poses that human beings are biologically and ecologically predisposed to engage in cultural and rational norms that regulate cooperation and structurally complex social projects (Andrews, 2015; Mameli, 2001; McGeer, 2007, 2015; Andrews, 2009; Zawidzki, 2008, 2013). From this perspective, our behavioral predictions are based on expectations generated by norms and behavioral routines we acquired in our social niche. We behave appropriately in social contexts because we have been taught to do so; we predict what the others will do in social contexts because they have been taught to behave according to norms and patterns we recognize. Our mind has been shaped for navigating social situations.

The mindshaping view claims that the centrality of social cognition relies on regulating our behavior according to those norms and routines that structure our interactions. Thus, coordination, mindreading and other types of socio-cognitive capacities rest on mindshaping mechanisms of norms and acquisition of behavioral routines. Furthermore, contrary to the representational assumption, the mindshaping view makes expectations rest on those social and rational norms regulating our social exchanges. Of course, the mindshaping view recognizes the use of attributing mental states in our socio-cognitive practices. However, this role is more restrictive. According to the defenders of this position, our folk psychological ascriptions have the role of normalizing or justifying behaviors that contravene the normative standard regulating our expectations. As Andrews (2012) puts it:

> Our understanding of others' beliefs and desires derives from a more basic understanding of others as intentional agents. We think about beliefs in particular kinds of situations, such as when a person deviates from expected behavior or violates the norms of society, but we don't need to appeal to beliefs to predict quotidian behavior. (Andrews, 2012, 10)

We recruit our mindreading capacities in very particular contexts, namely, those where our normative based anticipatory capacities fail to account for a certain behavior (For a similar point, see Gallagher & Hutto, 2008, 30, Millikan, 2004, 22, Zawidzki, 2013, 30)

Now, consider the examples we saw in Sect. 6.2.1 from the mindshaping viewpoint. In the hunting scenario, rather than postulating mental states, the Hunter_1's expectations concerning the prey are generated by general norms regarding how agents respond correctly to certain situations based on normative constrains of rationality. In other words, Hunter_1 expects the prey to flee beyond his position because he has learned that it is appropriate for agents to flee in a situation of danger.[3] A similar analysis can account for the Pub example. Norms and rituals

[3]One may object that understanding those norms of rationality requires the attribution of mental states. However, there is a more deflationist interpretation of those kinds of situations. Interpreting a particular behavior as being a rational only requires understanding an action as aiming a specific goal and constituting the most rational means given the environmental constraints. For different versions of this deflationist interpretation of behavior are the non-mentalistic version of "the

provide injunctions about what to do and expect. Making sense of the Pub situation involves a background of normative practices about economic transactions or costumer/service interactions. Nevertheless, the moral is the same: representing others' mental states is not required to solve the social interaction. Apart from rationality assumptions, we anticipate others' behavior in terms of stereotypes or social roles (Locksley, Borgida, Brekke, & Hepburn, 1980), generalizations over past behaviors (Kalish, 2002) or cultural norms (Kalish, 1998; Perner & Roessler, 2010). Our anticipatory skills rely on assumptions of "normalized behavior", the attributor presupposes the target is going to behave according to different standards of culturally normalized actions.

A critical dimension of the mindshaping view consists in explaining how an agent acquires the capacity to engage in the rational and cultural norms that generate expectations about behavior. Zawidzki (2013) postulates a set of psychological mechanisms that enhance the acquisition of norms and patterns. Some examples of those mechanisms are: imitation, natural pedagogy, norm recognition and enforcement, or stereotype assumptions.[4] Let me introduce some examples of those mechanisms. Contrary to other apes, human offspring do not only imitate others' behavior but they compulsively "overimitate" adult's behavior with a high grade of fidelity. This overimitation facilitates the acquisition of patterns of behavior which in its turn promotes social interactions. Firstly, it enhances prediction because once we copy some courses of behavior in some specific contexts; others will expect us to behave that way in such contexts. Secondly, overimitation enhances cooperation because it helps to copy cooperative patterns of behavior.

Other examples of mindshaping are the mechanisms for *natural pedagogy* (Csibra & Gergely, 2009). Natural pedagogy is a group of mechanisms which helps the infant to acquire certain behaviors from their caregivers. For instance, babies understand eye contact with their caregivers as a signal to attend to the subsequent behavior. This mechanism reinforces learning of actions. The infants learn which action is in order in each situation. Moreover, there are some tendencies in the caregivers that strengthen learning capacities, for instance, exaggerating their behavior. This generates a dynamic of learning which makes the infant understand the patterns of behaviors without the necessity of a theory of mind. A case of mindshaping mechanism in adults is the tendency to punish counter-normative actions. Henrich et al. (2005) have demonstrated a cross-cultural tendency in humans to punish some non-cooperative or irrational behavior in spite of the cost. This mechanism helps to homogenize individual behaviors to make them more predictable and cooperative.

intentional stance" (Zawidzki, 2013), "the teleological stance" (Gergely & Csibra, 2003) or "the situation theory" (Perner & Roessler, 2010). See also Sect. 6.3.1.

[4]Mindshaping mechanisms are a heterogeneous class. They can vary according to different variables: the target, the model, etc. Notice, for instance, that the mindshaping mechanism can be implemented in the mind that is shaped or in the mind that shapes the other mind.

Summing up, Human beings behave according to certain rational and cultural norms and specific routines that facilitate prediction and cooperation. Thus, an important aspect of our social interactions requires teaching, learning and urging each other to follow those patterns and norms. Those claims, I argue in Sect. 6.4, are important to approach some problems in social robotics. However, before I proceed to explain why, I consider some arguments to favor the approach.

6.3 Some Arguments for the Mindshaping View

In this section, I present two arguments supporting the mindshaping view. The first argument tries to undermine the two underlying assumptions of the mindreading view. By considering certain social situations and empirical evidence, I make the case that humans usually exploit normative standards that generate expectations about how others behave. Furthermore, I pose that even in cases where mindreading is necessary to explain some actions, they require considering a background of normative standards (Andrews, 2012). Secondly, I will defend that the mindshaping view can account for what is called the problem of tractability (Zawidzki, 2008). Contrary to the standard view, the mindshaping view can reduce the grade of under-determination that arises in the context of attributing mental states. This argument is of particular importance for social robotics since it points out to some of the problems of computational costs for implementing cognitive capacities in robots (Brooks, 1999).

6.3.1 Prediction and Explanation Without Mindreading

In the last section, we emphasized the central claim of mindshaping: our predictive capacities in social contexts rely on expectations generated by rational and cultural norms. This perspective opens a different strategy to approach social situations. According to it, we can make predictions based on normative standards rather than on mental states. Those norms facilitate our interactions by anticipating what others will do basis on what they should do according to those norms. In other words, once those norms and routines are available to the interpreter by means of mindshaping, the interpreter can ascribe *oughts* to the target without speculating about the internal causes of her behavior. Maibom (2007) explains this point as follows:

> Consider how people behave in restaurants. What the person who waits on guests does, he does qua waiter; his desire to take somebody's order is a function of him seeing himself as a waiter and is quite independent of his personal desires and preferences generally. When, perusing the menu, the other person sees him approaching, she infers that he is coming to take her order, but to do so she need only understand that this is what waiters do with customers and that he is a waiter and she is a customer. What he, personally, desires it is

irrelevant to the customer's prediction of what he will do and what she ought to do and vice versa. (Maibom, 2007, 568)

Those kinds of everyday explanations are difficult to understand from the mindreading view. Consider how difficult it would be to perform the interaction that Maibom proposed whether our behavior in restaurants had to be carried out by theorizing about mental states.

We exploit very different normative structures in order to generate those expectations. Including norms concerning situations, social roles, gender, rational patterns and so on. The diversity of strategies we find in human social skills is related to the different kinds of norms policing our social interactions. Consider the case of stereotypes for instance. Humans categorize other agents depending on various stereotypes concerning social roles, gender or ethnicity (Greenwald, McGhee, & Schwartz, 1998; Krueger & Clement, 1996; Olivola & Todorov, 2010). They exploit this categorization to generate expectations regarding others' behavior. In the case of gender, people expect males and females to behave differently. For example, female babies are expected to be more vulnerable in some situations than male babies, and we tend to associate different roles with each gender (Golombok & Fivush, 1994). In other words, stereotypes involve what Kalish and Lawson (2008) called "deontic relations": information about what a person should do or be like depending on how we categorize them. Our anticipatory capacities rely on the set of *oughts* we expect someone would undertake given these norms policing social interactions. It deserves mention that those social categories are not based on general inductive knowledge. We do not treat males and females differently because of differences in behavior, on the contrary we treat them as we do because we assume they should behave according to gender categories.

Similar *oughts* derive from the circumstances surrounding the agents on the basis of rational norms. For instance, when the interpreter perceives a building on fire, she can expect others to run away from the building. Those predictions do not necessitate desire or believe attributions. On the contrary, they rest on general rational constraints concerning what is rational or normal to do in a risky situation. This is what Gergely and Csibra (2003) calls the "teleological stance", "represent[ing] actions by relating relevant aspects of reality (action, goal-state and situational constraints) through the principle of rational action, which assumes that actions function to realize goal-states by the most efficient means available" (Gergely & Csibra, 2003, 287) (See also Zawidzki, 2013, 15). This capacity is exhibited in children far before they show full-fledged propositional ascriptions capacities, around the first year of life.[5]

[5]Traditionally, the appearance of full-fledged mental states ascriptions appears around four (Wellman, Cross, & Watson, 2001). However, recent studies point out to a far shorter age, around 12 months (Onishi & Baillargeon, 2005). However, those studies are subjected to more deflationist interpretations. Namely, considering that children can pass the tests thank the teleological stance or a similar low-level capacity (see Rakoczy, 2015, for a discussion).

What those cases emphasizes is that humans can exploit different normative schemas and routines to generate expectations about others' behavior. This idea confronts the assumption that the unique human capacity to understand others depends on the ability to represent their mental states. Furthermore, it casts into doubt the centrality that the mindreading view has given to mental states ascriptions. Another way to see this point is through the exploration of contexts where we usually ascribe mental states. As Andrews (2012) argues, attributing mental states is usually necessary only when our expectations are not fulfilled. Explaining others' behavior is only one of the many responses we deploy to deal with contra-normative behavior. When our expectations are violated, we respond by blaming, asking for justification or reprimanding for repairing our confidence in normative parameters (McGeer, 2007). Consider the following example; you see your colleague, who you always have considered a respectable person, yelling out in the middle of the street, creeping out people. Only in cases where your expectations are contravened, it makes sense to attempt to wonder what the target has in mind. Those cases of anomalous behaviors highlight the normative components of social cognition: we know that our expectations are grounded on norms because we react to the violation of those expectations with responses that reinforce them. Our mentalizing explanations are subsidiary of our regular normative practices. Only when those normatively based expectations fail, explanation appears on the stage. This way of understanding mentalizing would explain marking mental states in explanation is bounded to very specific cases of justification where we can make others to appears under a positives (Malle, 2004). As Malle has demonstrated, those cases represent a 13.6% of the cases when we are the actors, and about 17.6% of the observers.

In conclusion, there is an asymmetry between explanation and prediction in our social practices (Andrews, 2012) that demonstrates that mental state attribution is not as common as the mindreading view has supposed. Furthermore, we can anticipate others' behavior under rational and social normative standards without explaining what they are doing. Thus, we do not need to postulate mental entities for navigating social interactions. Now, we can see why the centrality assumption is misguided. Mindreading view has taken uncommon situations of justification as the paradigm for analyzing the majority of our socio-cognitive capacities. We do not mentalize others as often as the mindreading view supposes.

Before considering the next argument, I would like to face a possible counterargument for the view I am defending here. This counterargument would be as follows: Social interaction is normatively mediated. We need to follow norms and understand others as following them in order to tackle social interactions. However, understanding norms requires mindreading. To ascribe the capacity of following a rule to someone requires attributing him an intentional action, that is, an action caused by a belief/desire. Therefore, mindshaping presupposes mindreading capacities.

There are at least two lines of reply for this argument. Firstly, it is plausible to argue that a fundamental normative attitude or sensibility is logically independent of mindreading. Ginsborg (2011) has argued for a basic normative capacity that does not need mindreading abilities: the sense of appropriateness when engaging in a correct action. However, I would like to consider another compelling manner to face the argument. Andrews (2009) has provided empirical evidence that points out to the normative capabilities of animals to which it is doubtful we can attribute mindreading. For instance, chimpanzees exhibit the capacity to predict behaviors; they can deal with quite complex social environments, for example, a group with many members with different hierarchical status and a diverse set of behaviors according to that status. Moreover, they have some cultural traditions with a high grade of variability. Those practices seem to require understanding norms, at least implicitly. However, chimpanzees do not succeed in most of the standard tests for mindreading (Call & Tomasello, 1999). Furthermore, the sensitivity to norms has been discovered in very young children (Rakoczy, Warneken, & Tomasello, 2008) far before they succeed in those tests. Therefore, social animals without mindreading exhibit normative understanding, in other words, we do not have reasons to suppose that norm sensitivity requires mindreading abilities.

6.3.2 The Problem of Tractability

Another argument against the mindreading view is what Zawidzki (2013) calls *the problem of tractability*. A given action may be brought out by many different mental states. For example, believing we are late, believing we are missing the bus or wanting to reach a friend are all compatible with running from your work. This impossibility for a one to one match between mental states and behavior produces a high grade of indeterminacy for interpreting social situations. The problem is more pressing if we take into account that the interpretation of the agent should be sensitive to the context and the psychological background of the attributed person. A belief can motivate a particular course of behavior that otherwise can be contained by other mental states on the psychological background. Taking this into account, the interpretative problem of finding the inner cause of an action would be almost impossible. The computational cost of having an internal model of others' minds along with a mechanism for reducing this indeterminacy would be excessive from an evolutionary point of view.

In contrast, the mindshaping view presents a way to mitigate such indeterminacy. Agents are mutually interpretable because they behave according to the same cultural patterns and rational norms. The computational cost of interpretation is "outsourced" to the routines and norms governing social interactions. That restricts the grade of indeterminacy by making agents behave in a similar way. In this sense, social situations are analogous to other norm based practices. For instance, driving according to traffic norms. Car drivers do not need to calculate other drivers' behaviors in mentalistic terms because their behavior is regulated by traffic norms. The mindshaping view helps to make indeterminacy tractable in a way that is not available for the mindreading view. Moreover, this view seems to be more evolutionarily plausible. Evolutionary processes do not find optimal solutions to every problem. In fact, they usually use cheap solutions to costly problems (Dennett,

1996). Thus, from an evolutionary point of view, it is more reasonable to think that evolution would avoid using a costly expensive mechanism for representing other mental states and deal with a high grade of indeterminacy. In contrast, evolutionary processes could deal with the problem of coordination in humans by using capacities the brain already has, for instance, the capacity of matching patterns, mimicking, implicit learning, etc.

Notice that this problem is analogous to the problem of computational cost in classical artificial intelligence. As Brooks (1999) says:

> Over the years within traditional Artificial Intelligence, it has become accepted that they will need an objective model of the world with individuated entities, tracked and identified over time – the models of knowledge representation that have been developed expect and require such a one-to-one correspondence between the world and the agent's representation of it. (Brooks, 1999, 166)

The necessity of an internal model to solve cognitive problems has been the cornerstone of classical artificial intelligence and cognitive sciences. The problem with this model is that natural environments are complex. This complexity obligates the designer to decide between creating a highly abstract inner model or facing a high computational cost to internalize the complex nature of the environment. Contrary to this view, Brooks decided to build situated agents who must deal with the constraint of the real environment in real time. Instead of using internal representation, they use cheap and efficient tricks for generating situation-appropriate action, the best way to act according to specific goals. Brooks's argument illustrates the computational problem of classical artificial intelligence to deal with complex environments. If we transfer this issue to social cognition, we see why implementing an internal model of others' minds is costly. However, the mindshaping view presents some computationally cheap solutions to social problems. Instead of implementing this model in robots, we can implement some mechanisms that regulate their behavior according to some standards and other robots' expectations, reproducing the same dynamics that we saw in Sect. 6.2.2.

6.4 Why Mindshaping Matters for Social Robotics

At this point, it is not difficult to perceive how the mindshaping view can contribute to thinking about social robotics. On the one hand, characterizing social interactions from a different angle facilitates re-thinking some methodologies of social robotics. On the other hand, this view offers a manner to avoid the computational cost of implementing an internal model of others' minds. Moreover, I believe there are specific problems and areas of social robotics which would be especially interesting to approach from the mindshaping perspective. Those are developmental social robotics, the relationship between social robotics and general intelligence robotics, and the development of companion robots.

6.4.1 Developmental Robotics

A considerable scope of application for social robotics is scientific research on social cognition. Firstly, we can explore theories of mind by implementing functional analogues of socio-cognitive mechanisms in robots. As Breazeal (2002) puts it:

> Robots could be used as experimental testbeds for scientific inquiry. Computational models of our social abilities could be implemented, tested, and analyzed on robots as they participate in controlled social scenarios. In this way, robots could potentially be used in the same studies and experiments that scientists use to understand human social behavior. Robot data could be compared with human performance under similar conditions. Differences between the two could be used to refine the models and inspire new experiments. (Breazeal, 2002, 1)

Secondly, social robotics can explore different approaches of the mind in order to improve the building of mechanical agents. *Developmental robotics* exemplify those two ways of interrelating the study of the mind with robotics. Developmental robotics is an area of research in the intersection between social robotics and developmental psychology. As Lungarella, Metta, Pfeifer, and Sandini (2004) express it: "it constitutes an interdisciplinary and two-pronged approach to robotics, which on one side employs robots to instantiate and investigate models originating from developmental sciences, and on the other side seeks to design better robotic systems by applying insights gained from studies on ontogenetic development" (Lungarella et al., 2004, 151). There are two motivations behind this area of research: (1) Artificial intelligence's necessity of developing new methodologies and models in the construction of social robots. (2) With robots developmental psychology and neuroscience can test different developmental models of cognition, which they otherwise cannot.

Regarding the first motivation, an important aspect of mindshaping is its emphasis in the ontogeny of social cognition. If navigating the social world demands mindshaping mechanisms, they would provide social robotics with different tools to develop robots. For instance, equipping robots with capacities that enhance acquisition of norms and patterns, in contrast with providing them with models of the others' mind. Equipping robots with those mechanisms implies making them more sensitive to instructions or reinforcement of behavior, for example. Furthermore, those tools would help to reduce the computational costs, even if implementing internal models is necessary. For instance, equipping a robot with a mechanism that reinforces some actions when the user rewards them, it would automatize the behavior in the context. The robot would not have to solve the problem every time it confronts the context. In this respect, we can see some studies that remind the mindshaping methodology, for instance, robots equipped with learning mechanisms guided by robot-directed speech (Breazeal & Aryananda, 2002). In those studies, the users can directly modulate the robot's affective states through verbal communication. The robot is implemented with a system whereby a humanoid robot recognizes and affectively responds to praise, prohibition, attention, and comfort in robot-directed speech. Those communicative intents were used to direct the robot's attention to the relevant aspects of the task.

In principle, these studies are not incompatible with the mindreading view. However, notice that the communicative intent recognition is more than a mechanism that provides the mindreading system with social information for further inferences. The communicative intent recognition mechanism shapes the robot capacity to perform a task without the necessity of implementing complex inner models. In this sense, this study respects the spirit of the mindshaping view. Another instance of how developmental psychology may inspire social robotics is the capacity known as joint attention. Joint attention is the ability of selective attending to an object of mutual interest. This ability is considered the precursor of high-level social capacities such as mindreading. Scassellati (1999) carries out a thorough study of how joint attention can be implemented in robots. He discomposes the ability into four subtasks that the robot must perform: maintaining eye contact, gaze following, imperative pointing and declarative pointing. The significant difference between the mindshaping view and the mindreading view is how to consider the role of joint attention in social cognition. According to the mindreading view, joint attention is a mechanism that provides mindreading capacities with social cues to infer beliefs and desires. On the other hand, the mindshaping view considers joint attention not only as a mechanism for gaining information but also for reinforcing and learning patterns of behavior. As we said in Sect. 6.2.2, Csibra and Gergely (2009) defend that infants interpret eye contact as a clue to pay attention to the subsequent behavior the caregivers perform. Following this line, social robotics can equip robots with joint attention structures, like imperative pointing, which reinforce conducts or sign which behavior should be attended for mimicking the target. In this sense, joint attention is a mindshaping mechanism, a mechanism for learning the appropriate manner to behave.

On the other hand, implementing social robots can help us to understand our social skills. The mindshaping view gives us a substantive theory about what types of mechanisms underlie social cognition. We must test whether or not postulating mindshaping mechanisms of different types helps to solve paradigmatic social contexts. The existence and efficacy of those mechanisms are empirical questions to which developmental robotics can shed some light. Social robotics can attempt to implement those mechanisms to find how they work. Also, social robotics can test the viability of both competing models. For instance, demonstrating if social robots modeled through the mindshaping view reproduce human patterns of social cognition more accurately than those modeled through the mindreading view. In this sense, the mindshaping view would predict that a robot with mindshaping mechanisms would simulate human social interactions more accurately than a robot equipped with mindreading capacities.

6.4.2 General Intelligence and Social Robotics

Providing models for developing robots is not only a necessity of social robotics in particular, but also a necessity for general intelligence robotics as well. The relationship between social and non-social intelligence affects the development of general intelligent robots. As Dautenhahn (2007) argues, if the social intelligence hypothesis is correct and the evolution of our general intelligence is tied to social skills, then "developing an intelligent robot means developing first a socially intelligent robot" (Dautenhahn, 2007, 682). The standard hypothesis about human intelligence evolution is the social intelligence hypothesis according to which human intelligence evolved due to evolutionary pressures of the social complexity. If this hypothesis is correct, a reliable manner to build intelligent robots can be to copy the strategy of evolution. In other words, building intelligent robots should be based on building social robots.

Although there are several versions of the social intelligence hypothesis, what Dauthenhan has in mind is the Machiavellian intelligence hypothesis (Byrne & Whiten, 1988). According to this view, primate intelligence evolved as an adaptation to deal with the social complexity: manipulation, prediction and explanation of conspecifics. The rationale behind this view is that our social intelligence evolved in order to solve the problem of the cheater: in a social group where every member cooperates, a cheater who takes advantage of the other members without paying any cost would break the social dynamic of the group. Therefore, a mindreading mechanism would help us to detect those cheaters. In other words, when social exigencies started to work as evolutionary pressure, they provide us with bigger brains and mindreading capacities. Then, some of those abilities were transferred to non-social intelligence. Social intelligence influenced non-social intelligence (Gigerenzer, 1997). Using the term introduced by Gould and Vrba (1982), non-social intelligence was an *exaptation* for social intelligence, that is, although our intelligence was adapted to solve social problems, it was co-opted to address non-social problems. The implication of this view for social robotics is that non-social intelligence takes advantage of the knowledge, heuristics and routines of social skills.

In contrast to the Machiavellian intelligence hypothesis, the mindshaping perspective presents a different manner to think about the evolution of human social and general skills. The mindshaping approximation offers a more cooperative view of human evolution. According to it, the cheater problem has been overestimated, since the cooperation was an easy solution for surviving in the evolutionary environment of our ancestors:

> But in contrast to the usual defection-management perspective, I see the key cognitive preconditions of cooperation as being those necessary for coordination and effective joint action. In a complex and risky environment, profitable cooperation depends on more than trust, on more than the confidence that you will not be ripped off. It also depends on teamwork, with a mutual awareness of one another's abilities, and on integrating this social information with appropriate information about the world: the target of joint action. (Sterelny, 2013)

This cooperative environment[6] provoked the dynamics necessary for the emergence of group selection. In the contexts of group selection, social learning was a useful tool for surviving. Teaching and learning hunting strategies, folk biology, knowledge of the environment and different strategies to reason about general problems were central linchpins to motivate the group dynamics. Keeping that in mind, we can see the contrast with the Machiavellian intelligence hypothesis. Although both views emphasize the social origin of human intelligence, the mindshaping view bases human intelligence on the capacity to learn different ways to approach problems in the world facilitated for mindshaping capacities. As a cooperative animal, we cooperate to solve problems together. Once the social dynamics appears, the human niche serves as social scaffolding for our general intelligence. Being taught to behave rationally does not mean only to behave rationally in our interaction with others but being trained to deal rationally with the world. For instance, when humans learn to solve general problems like designing cooperative strategies, building tools or recognizing appropriate information for a given goal, they do not have to apply necessarily the same kind of strategies they use to understand others' behaviors, but to learn norms relative to those problems in social contexts. The social niche is a pedagogical environment for acquiring general skills.

Now, the question is how those differences in understanding human evolution can be translated to differences in social robotics. Designing intelligent robots necessitate creating robots able to interact with us in the world. Those robots should be sensitive not only to our responses and expressions but to take those expressions and responses as signals to learn how to approach non-social problems. Those exchanges would provoke the dynamic interactions between humans and robots naturally arise. Our social practices are situated in cooperative practices where we face the environment together. Social robotics should try to replicate such practices in the human-robot interactions. A good example could be the joint attention capacity applied to mimic. Implementing a joint attention mechanism would facilitate the mimicking of the appropriate courses of actions in each context. In this case, the mechanism uses social context to deal with environments, instead of a social mechanism recruited for a non-social purpose. However, consider another example we can find in social robotics literature and how it exemplifies this manner of understanding the relationship between social and non-social cognition. Brooks, Breazeal, Marjanovic, Scassellati, and Williamson (1999) describe the details of the Cog Project, which tries to build humanoid robots.

An interesting point of this paper is the role that Brooks and his colleagues assign to the emotional system of robots:

[6]This is a simplified version of the thesis. Zawidzki defends the idea that those environments enhanced cooperation of our ancestors. However, as soon as the environments started changing, the population began a process of balkanization which exerts a selective pressure for the emergence of mindshaping and group selection (see Zawidzki, 2013, Ch. 4 for more details).

> An important function for a robot's motivational system is not only to establish appropriate interactions with the caretaker, but also to regulate their intensity so that the robot is neither overwhelmed nor under stimulated by them. When designed properly, the intensity of the robot's expressions provides appropriate cues for the caretaker to increase the intensity of the interaction, tone it down, or maintain it at the current level. By doing so, both parties can modify their own behaviour and the behaviour of the other to maintain the intensity of the interaction that the robot requires. (Brooks et al., 1999, 66)

Notice that this emotional system is described as a mindshaping mechanism. The emotional system helps to facilitate learning during social exchanges. In this sense, the emotional system can be used as an external reward signal that reinforces good actions and penalizes the bad ones. The relationship between robots and users is parallel to the relationship between parents and infants: "parents take an active role in shaping and guiding how and what infants learn by means of *scaffolding*. As the word implies, the parent provides a supportive framework for the infant's interactions with the environments to foster novel abilities" (Brooks et al., 1999, 66). Given this, we can see the contrast with the Machiavellian perspective, where the socio-cognitive capacities arise for social purposes, and then, they are recruited for solving general problems. The alternative presents a set of skills that prepare our minds to acquire novel abilities for general purposes. Moreover, those systems improve learning in more natural contexts and give a high range of behaviors to learn. The robot could be allowed to learn from its own actions and experiences while guided by humans; then, the set of behaviors and tasks able to learn is higher (Gu & Hu, 2004; Nicolescu, Natural methods for learning and generalization in human-robot domains. Unpublished doctoral dissertation. University of Southern California, Computer Science Department, 2002).

6.4.3 Companion Robots

The third field where I think the mindshaping view could help to take a different perspective is companion robots. A companion robot is a robot whose purpose is to serve humans as assistants or "companions" (Cogniron, 2007):

> The overall objectives of this project are to study the perceptual, representational, reasoning and learning capabilities of embodied robots in human centred environments. The project develops methods and technologies for the construction of such cognitive robots, able to evolve and grow their capacities in close interaction with humans in an open-ended fashion. (Cogniron, 2007)

Developing a companion robot requires two dimensions or points of view: the human-centered view and the robot cognition viewpoint (Dautenhahn, 2007). According to the first dimension, the companion robot should be able to deal with social interactions with humans involved. Those robots should be sensitive to human necessities and expectations and respond according to them. Breazeal (2005) makes explicit her worries about the need of building such sensitive robots in that way:

> [Actual] robots are not really aware of our goals and intentions. As a result, they don't know how to appropriately adjust their behavior to help us as our goals and needs change. They generally do not flexibly draw their attention to what we currently find of interest so that their behavior can be coordinated and information can be focused about the same thing. (Breazeal, 2005, 19)

We must build robots with the capacity of "understanding" human behavior. According to the second dimension, our companion robot should be able to interact with the world: carrying out different tasks and solving problems for facilitating human life.

Robots with mindshaping mechanisms should learn to perform projects according to the needs and expectations of humans. As Dautenhan says, "A truly personalized robot companion takes into consideration an individual human's likes, dislikes and preferences and adapts its behaviour accordingly" (Dautenhahn, 2007, 685). However, this does not require implementing an internal model of human minds in the robot. As we saw in Sect. 6.2.2, social abilities require being sensitive to norms and patterns of behavior. Thus, implementing mindshaping mechanisms in robots can help to build robots able to interact according to those norms. Notice that the connection between those two dimensions is quite similar to the connection between social cognition and non-social cognition presented in the last section. Therefore, we can see how the mechanisms we presented before may improve companion robots. As we explained before, equipping robots with joint attention, emotional system or speech recognition can enhance the learnability of the robot substantially. In the context of companion robots, this could provide the robot with more friendly attitude and personalized treatment. Being shaped by his users, the robot could be sensitive to individual idiosyncratic behaviors and expectations. Moreover, the capacity of the robots to react emotionally and regulate its behavior according to the user can help to avoid partially the "uncanny valley" effect, which is significant for companion robots.

6.5 Conclusions

In this paper, I have considered how the mindshaping framework in the study of social cognition can help to rethink some areas of social robotics. According to the mindshaping view, our socio-cognitive abilities have a significant normative aspect related to specific cultural and rational norms, learning and teaching capacities and reinforcement of those norms. This normative dimension and some other features of mindshaping are significant for social robotics. Firstly, mindshaping is important for developmental robotics due to its emphasis on the developmental aspect of social cognition. Secondly, it supposes an alternative to the Machiavellian intelligence hypothesis, and thus, it can shed light on the relationship between general and social intelligence. Finally, I regard mindshaping mechanisms as providing a new way to look at companion robotics. Companion robots need to perform actions in the world according to the needs of humans. In other words, they need mechanisms to regulate their own behavior according to human expectations.

Acknowledgements The author wishes to thank the comments by Fernando Martinez Manrique, Johanna Seibt, Raul Hakli, the Ph.D. students of the Philosophy Department 1 at Universidad de Granada, Charlie White and the audience at the Robo-Philosophy Conference 2014. Research for this paper was funded by the Spanish Government through Research Projects FFI2015-65953-P and the fellowship FPI BES-2012-052157.

References

Andrews, K. (2009). Understanding norms without a theory of mind. *Inquiry, 52*(5), 433–448.
Andrews, K. (2012). *Do apes read minds? Toward a new folk psychology*. Cambridge, MA: MIT Press.
Andrews, K. (2015). The folk psychology spiral. *Southern Journal of Philosophy, 53*(S1), 50–67.
Baron-Cohen, S. (1994). How to build a baby that can read minds: Cognitive mechanisms in mindreading. *Current Psychology of Cognition, 13*, 513–552.
Baron-Cohen, S. (1999). The evolution of a theory of mind. In M. Corballis & S. Lea (Eds.), *The descent of mind*. New York: Oxford University Press.
Breazeal, C. (2002). *Designing Sociable Robots*. Cambridge: MIT Press.
Breazeal, C. (2003). Towards sociable robots. *Robotics and Autonomous Systems, 42*, 167–175.
Breazeal, C. (2005). Socially intelligent robots. *Interactions, 12*(2), 19–22.
Breazeal, C., & Aryananda, L. (2002). Recognition of affective communicative intent in robot-directed speech. *Autonomous Robots, 12*, 83–104.
Brooks, R. (1999). *Cambrian intelligence: The early history of the new AI*. Cambridge: MIT Press.
Brooks, R., Breazeal, C., Marjanovic, M., Scassellati, B., & Williamson, M. (1999). The Cog project: Building a humanoid robot. In C. Nehaviv (Ed.), *Computation for metaphors, analogy and agents* (pp. 176–195). Berlin/New York: Springer.
Byrne, R. W., & Whiten, A. (Eds.). (1988). *Machiavellian intelligence: Social expertise and the evolution of intellect in monkeys, apes and humans*. Oxford: Oxford University Press.
Cabibihan, J.-J., Javed, H., Ang Jr M., & Aljunied, S. M. (2013). Why robots? A survey on the roles and benefits of social robots in the therapy of children with autism. *International Journal of Social Robotics, 5*(4), 593–618.
Call, J., & Tomasello, M. (1999). A nonverbal false belief task: The performance of children and great apes. *Child Development, 70*(2), 381–395.
Carruthers, P. (1996). Simulation and self-knowledge: A defence of theory-theory. In P. Carruthers & P. Smith (Eds.), *Theories of theories of mind* (pp. 22–38). Cambridge: Cambridge University Press.
Cogniron. (2007). *The cognitive robot companion*. Retrieved from http://www.cogniron.org (LAAS-CNRS, Toulouse, France. An Integrated Project funded by the European Commission's Sixth Framework Programme, Accessed 14 Sep 2016)
Csibra, G., & Gergely, G. (2009). Natural pedagogy. *Trends in Cognitive Science, 13*, 148–153.
Dautenhahn, K. (2007). Socially intelligent robots: Dimensions of human-robot interaction. *Philosophical Transactions of the Royal Society B, 362*, 679–704.
Dennett, D. C. (1996). *Darwin's dangerous idea: Evolution and the meanings of life*. New York: Simon & Schuster.
Dunbar, R. (2000). On the origin of the human mind. In P. Carruthers & A. Cham-berlain (Eds.), *Evolution and the modern mind: Modularity, language and meta-cognition* (pp. 238–253). Cambridge: Cambridge University Press.

Dunbar, R. (2003). The social brain: Mind and language, and society in evolutionary perspective. *Annual Review of Anthropology, 32*, 163–181.
Fodor, J. A. (1987). *Psychosemantics: The problem of meaning in the philosophy of mind*. Cambridge, MA: MIT Press.
Fong, T., Nourbakhsh, I., & Dautenhahn, K. (2003). A survey of socially interactive robots. *Robotics and Autonomous Systems, 42*, 143–166.
Forguson, L., & Gopnik, A. (1988). The ontogeny of common sense. In J. W. Astington, P. L. Harris, & D. R. Olson (Eds.), *Developing theories of mind* (pp. 226–243). Cambridge: Cambridge University Press.
Gallagher, S. (2001). The practice of mind: Theory simulation or primary interaction? *Journal of Consciousness Studies, 8*(5–6), 83–108.
Gallagher, S. (2004). Understanding interpersonal problems in autism: Interaction theory as an alternative to theory of mind. *Philosophy Psychiatry Psychology, 11*, 199–217.
Gallagher, S., & Hutto, D. D. (2008). Understanding others through primary interaction and narrative practice. In J. Zlatev, T. Racine, C. Sinha, & E. Itkonen (Eds.), *The shared mind: Perspectives on intersubjectivity* (pp. 17–38). Amsterdam/Philadelphia: John Benjamins.
Gergely, G., & Csibra, G. (2003). Teleological reasoning in infancy: The naïve theory of rational action. *Trends in Cognitive Sciences, 7*, 287–292.
Gigerenzer, G. (1997). The modularity of social intelligence. In A. Whiten & H. Byrne (Eds.), *Machiavellian intelligence II: Extensions and evaluations* (pp. 264–288). Cambridge, UK: Cambridge University Press.
Ginsborg, H. (2011). Primitive normativity and skepticism about rules. *The Journal of Philosophy, CVIII*(5), 227–254.
Goldman, A. I. (2006). *Simulating minds: The philosophy, psychology and neuroscience of mindreading*. New York: Oxford University Press.
Golombok, S., & Fivush, R. (1994). *Gender development*. New York: Cambridge University Press.
Gordon, R. M. (1996). Radical simulationism. In P. Carruthers & P. Smith (Eds.), *Theories of theories of mind* (pp. 11–21). Cambridge: Cambridge University Press.
Gould, S. J., & Vrba, E. S. (1982). Exaptation—a missing term in the science of form. *Paleobiology, 8*(1), 4–15.
Greenwald, A. G., McGhee, D. E., & Schwartz, J. L. K. (1998). Measuring individual differences in implicit cognition: The implicit association test. *Journal of Personality and Social Psychology, 74*, 1464–1480.
Gu, D., & Hu, H. (2004). Teaching robots to coordinate its behaviours. In *Proceedings of IEEE International Conference on Robotics and Automation (ICRA'04)* (Vol. 4, pp. 3721–3726). New Orleans, U.S.A.
Heal, J. (2005). Joint attention and understanding the mind. In J. Roessler (Ed.), *Joint attention: Communication and other minds* (pp. 34–44). Oxford: Oxford University Press.
Henrich, J., Boyd, R., Bowles, S., Camerer, C., Fehr, E., Gintis, H., … others (2005). "Economic man" in cross-cultural perspective: Behavioral experiments in 15 small-scale societies. *Behavioral and Brain Sciences, 28*(6), 795–855.
Hutto, D. D., & Ratcliffe, M. (Eds.). (2007). *Folk psychology re-assessed*. Dordrecht, The Netherlands: Springer.
Kalish, C. W. (1998). Reasons and causes: Children's understanding of conformity to social rules and physical laws. *Child Development, 69*, 706–720.
Kalish, C. W. (2002). Children's predictions of consistency in people's actions. *Cognition, 84*, 237–265.
Kalish, C. W., & Lawson, C. A. (2008). Development of social category representations: Early appreciation of roles and deontic relations. *Child Development, 79*, 577–593.
Krueger, J., & Clement, R. W. (1996). Inferring category characteristics from sample characteristics: Inductive reasoning and social projection. *Journal of Experimental Psychology: General, 125*, 52–68.
Leudar, I., & Costall, A. (Eds.). (2009). *Against theory of mind*. Basingstoke [England]/New York: Palgrave McMillan.

Locksley, A., Borgida, E., Brekke, N., & Hepburn, C. (1980). Sex stereotypes and social judgment. *Journal of Personality and Social Psychology, 39*, 821–831.

Lungarella, M., Metta, G., Pfeifer, R., & Sandini, G. (2004). Developmental robotics: A survey *Connection Sciences, 15*, 151–190.

Maibom, H. L. (2007). Social systems. *Philosophical Psychology, 20*, 557–578.

Malle, B. F. (2004). *How the mind explains behavior: Folk explanations, meaning, and social interaction*. Cambridge: MIT Press.

Mameli, M. (2001). Mindreading, mindshaping, and evolution. *Biology and Philosophy, 16*(5), 595–626.

McGeer, V. (2007). The regulative dimension of folk psychology. In D. D. Hutto & M. Ratcliffe (Eds.), *Folk psychology re-assessed* (pp. 138–156). Dordrecht: Springer.

McGeer, V. (2015). Mind-making practices: The social infrastructure of self-knowing agency and responsibility. *Philosophical Explorations, 18*(2), 259–281.

Millikan, R. G. (2004). *Varieties of meaning: The 2002 Jean Nicod lectures*. Cambridge, MA: MIT Press.

Mithen, S. (1996). *The pre-history of the mind: A search for the origins of art, religion and science*. London: Thames and Hudson.

Moral, S., Pardo, D., & Angulo, C. (2009). Social robot paradigms: An overview. In *Proceedings of the 10th International Work Conference on Artificial Neural Networks: Part I: Bio-inspired Systems: Computational and Ambient Intelligence* (pp. 773–780). Salamanca, Spain.

Nichols, S., & Stich, S. (2003). *Mindreading: An integrated account of pretence, self-awareness, and understanding other minds*. Oxford: Oxford University Press.

Olivola, C., & Todorov, A. (2010). Elected in 100 milliseconds: Appearance-based trait inferences and voting. *Journal of Nonverbal Behaviour, 34*, 83–110.

Onishi, K. H., & Baillargeon, R. (2005). Do 15-month-old infants understand false beliefs? *Science, 308*(5719), 255–258.

Perner, J., & Roessler, J. (2010). Teleology and causal understanding in children's theory of mind. In J. Aguilar & A. Buckareff (Eds.), *Causing human actions: New perspectives on the causal theory of action* (pp. 199–228). Cambridge, MA: MIT Press.

Rakoczy, H. (2015). In defense of a developmental dogma: Children acquire propositional attitude folk psychology around age 4. *Synthese*, 1–19.

Rakoczy, H., Warneken, F., & Tomasello, M. (2008). The sources of normativity: Young children's awareness of the normative structure of games. *Develop- mental Psychology, 44*(3), 875–881.

Scassellati, B. (1999). Imitation and mechanisms of joint attention: A developmental structure for building social skills on a humanoid robot. In C. Ne- haviv (Ed.), *Computation for metaphors, analogy and agents* (pp. 176–195). Berlin/New York: Springer.

Sterelny, K. (2013). *The evolved apprentice: How evolution made humans unique*. Cambridge: MIT Press.

Wellman, H. M., Cross, D., & Watson, J. (2001). Meta-analysis of theory-of-mind development: The truth about false belief. *Child Development, 72*(3), 655–684.

Zawidzki, T. (2008). The function of folk psychology: Mind reading or mindshaping? *Philosophical Explorations, 11*(3), 193–209.

Zawidzki, T. (2013). *Mindshaping: A new framework for understanding human social cognition*. Cambridge: MIT Press.

Chapter 7
An Affordance-Based Conceptual Framework for Spatial Behavior of Social Robots

Felix Lindner and Carola Eschenbach

Abstract Socially aware robots have to coordinate their actions considering the spatial requirements of the humans with whom they interact. We propose a general framework based on the notion of affordances that generalizes geometrical accounts to the problem of human-aware placement of robot activities. The framework provides a conceptual instrument to take into account the heterogeneous abilities and affordances of humans, robots, and environmental entities. We discuss how knowledge about (socio-)spacial aspects of affordances can be used in various reasoning tasks relevant to human-robot interaction. Applying the notion of a practical reason, socially-aware robots are able to solve the social activity-placement problem.

Keywords Social robots • Spatial behavior • Conceptual model • Affordances

7.1 Introduction

Gibson (1977) introduced the notion of affordance as the central concept of ecological psychology. According to him, affordances are the possibilities for actions the environment offers to agents in cases where the abilities of agents and properties of the environment match. For example, chairs afford sitting to humans, stairs afford climbing to many humans, but form obstacles for many types of robots and for humans in wheelchairs, and some types of ramps afford climbing to both humans and robots. Both humans and robots also provide affordances to other humans and robots, thereby enabling interaction with other agents.

F. Lindner (✉)
Department of Computer Science, Foundations of Artificial Intelligence Group, Albert-Ludwigs-University of Freiburg, Freiburg, Germany
e-mail: lindner@informatik.uni-freiburg.de

C. Eschenbach
Department for Informatics, Knowledge and Language Processing Group, University of Hamburg, Hamburg, Germany
e-mail: eschenbach@informatik.uni-hamburg.de

R. Hakli, J. Seibt (eds.), *Sociality and Normativity for Robots*, Studies in the Philosophy of Sociality 9, DOI 10.1007/978-3-319-53133-5_7

The concept of affordance has inspired several researchers in various disciplines such as psychology (Turvey, 1992; Chemero, 2003; Stoffregen, 2003), human-computer interaction (Norman, 1999), geographic information systems (Kuhn, 2007; Raubal, 2001), and robotics (Stoytchev, 2008; Şahin, Çakmak, Doğar, Uğur, & Üçoluk, 2007; Saffiotti & Broxvall, 2008; Raubal & Moratz, 2008; Krüger et al., 2011), just to name a few. In philosophy, the notion of affordances has been related to phenomenology of Merleau-Ponty (Dohn, 2009). From the perspective of Gibson (1977), there is meaning in the world (i.e., affordances) and therefore agents can act. This is complementary to the view taken by some phenomenologists, viz., agents can act and therefore the world is meaningful (Dohn, 2009).

Traditionally, in robotics, the notion of affordances is related to the action possibilities of a single agent (e.g. Stoytchev, 2008). However, if agents have to coordinate their actions, taking care of the action possibilities of someone else becomes essential. Thus, the world is not just meaningful to a particular agent insofar as the world offers possibilities for action to this agent, but also insofar as it offers possibilities for action to other agents. Reasoning about affordances in social robotics requires to take agents with different abilities and needs into account. In this article, we put the focus on the spatial structure of affordances. This, we argue, is pivotal for the design of socially acceptable autonomous robot behavior.

We have presented a taxonomy of socially meaningful spaces (Lindner & Eschenbach, 2011; Lindner, 2015) including *affordance spaces* and *activity spaces* produced by affordances and activities, respectively.[1] In a nutshell, activity spaces are spaces used by agents while performing an activity, and affordance spaces are spaces that enable the execution of the afforded activity to an agent. (In the following, the term *activity* refers to any kind of event or process involving an agent, independently of the temporal aspect or abstraction level.)

In comparison to affordances, activities are short-lived. Activities exist only as long as they take place. Affordances are possibilities of activities and can exist independently of activities actually taking place. A single affordance can be realized more than once by different activities. The approach to affordances we present here also allows that different agents can act upon a single affordance.

Activity spaces of activities taking place simultaneously can yield conflicts, e.g., when two agents try to move through one door at the same time. Social interaction requires taking care of the spatial requirements of the other's activities and coordinating such activities. Similarly, if an affordance space is blocked, e.g., by a robot parking in some region, an agent may be unable to perform some activity. Thus, independently of whether there is an agent present that plans to act upon an affordance, social action planning should take care of affordance spaces for which it is to be expected that some agent will use it.

[1]In this context, "social space" refers to spatial structures that are significant for assessing the social appropriateness of agent behavior. Thus, the use of this notion is not limited to social spaces like the internet or public parks.

Taking affordances to be (mere) possibilities of activities with different agents, the spatial requirements of different agents acting upon an affordance can vary. While a small robot parking in a doorway might not hinder many humans to walk through, it might be an obstacle for a father with a baby-stroller. Thus, to act in a socially aware manner regarding space requires knowledge about other inhabitants of one's environment regarding their spatial requirements in acting. As these inhabitants might differ regarding their abilities and spatial requirements, a general model must allow for different affordance spaces for different types of agents.

The next section gives a short overview on affordances in robotics. Section 7.3 discusses activity spaces and affordance spaces in relation to each other and provides a coherent model of affordances and affordance spaces interrelating them with relevant concepts. Section 7.4 shows how knowledge about affordances and affordance spaces can be used for different spatial reasoning tasks and motivates the need for rating affordance spaces regarding social effects of their use within an environment. Section 7.5 introduces reasons related to affordance spaces and expected co-inhabitants of the environment as a means to identify socially relevant regions and address the problem of choosing a socially adequate place to carry out an activity. The conceptual framework is applied to two case studies presented in Sect. 7.6. Section 7.7 discusses the contributions of our framework and provides an outlook.

7.2 Affordances in Robotics

With a focus on affordance-based robot control, Şahin et al. (2007) identify three different perspectives for describing affordances. The observer perspective relates agents, environmental entities, behaviors, and effects. The agent perspective fixes an agent and relates the other three components, while the environmental perspective fixes an environmental entity.

The agent perspective is the perspective taken in approaches to single-robot learning and planning. An example for affordances in the agent perspective is given by Stoytchev (2008). He proposes a representation of affordances based on observations before and after a robot executes a behavior on an entity using a particular tool. That way, the robot learns the effects of behaviors and, thus, what the entity affords to it. As the derived representations are tightly coupled to the specific perceptual-motor-capabilities of the robot platform at hand, they cannot be used to reason about affordances the entity provides to other agents. Much work in this direction has also been done in robot manipulation planning (cf. Krüger et al., 2011).

If agents with diverse abilities cooperate, they must be able to reason about the affordances provided by and provided for other agents just as about their own. Saffiotti and Broxvall (2008) present an ecology of robots with heterogeneous abilities that can co-jointly solve tasks. Therefore, each robot publishes its functionalities to a central unit, which allows to integrate the functionalities available in the different robots and use a subset of these functionalities to plan a configuration that fits to the current context or task. However, the integration of humans in this approach is still an unsolved problem.

Some approaches concerned with spatial aspects of human-robot interaction address the question of how affordances structure space and yield spatial restrictions. For example, Yamaoka, Kanda, Ishiguro, and Hagita (2008) focus on how a robot with the task to present information about an object to a human should place itself relative to the listener and relativ to the presented object. To analyze this problem, the authors employ F-Formations, a spatial model of activities proposed by Kendon (1990). In Sect. 7.4, we will indicate how F-Formations can be derived from the combination of more simple affordance spaces.

Sisbot, Marin-Urias, Broquère, Sidobre, and Alami (2010) consider several constraints for human-aware placement planning that take both human properties and robot properties into account. For example, to determine a good pose for handing an object to a human, a robot should take the human's field of view into account as well as its own arm length.

The approaches of Yamaoka and colleagues, and of Sisbot and colleagues can be reframed in terms of socially adequate spatial behavior based on affordances of robots and their human interactants. However, a link to affordance theory has not been explicitly established by the authors.

A general approach to human-aware spatial behavior of robots requires to model the influence of affordances on the functional and social structure of space. The dominating agent perspective of affordances seems to result in neglecting the task of recognizing affordances the environment might provide for agents with diverging abilities and spatial constraints. For example, humans may be interested in viewing pictures or other kinds of displays. Even though these kinds of activity might not be relevant for robots, taking the perspective of humans in spatial planning also involves taking care of such human activities and their spatial requirements. Moreover, a more elaborated approach should consider the heterogeniety of robot platforms as well as the heterogeneity of human beings, which also differ regarding abilities and spatial requirements.

This short discussion of affordances shows that the spatial structure imprinted by affordances on the surrounding has only been considered for very specific interaction contexts. The social dimension deriving from the possibility that agents can deactivate affordances for other agents due to inappropriate behavior is largely ignored.

7.3 Activity Spaces and Affordance Spaces

7.3.1 General Structure of Activity Spaces

Activities are located in space (Kendon, 1990). Any activity occurs somewhere and has a relationship to the spatial regions the participants of the activity are located in. Kendon (1990) discusses so-called F-Formations, which are spatial structures

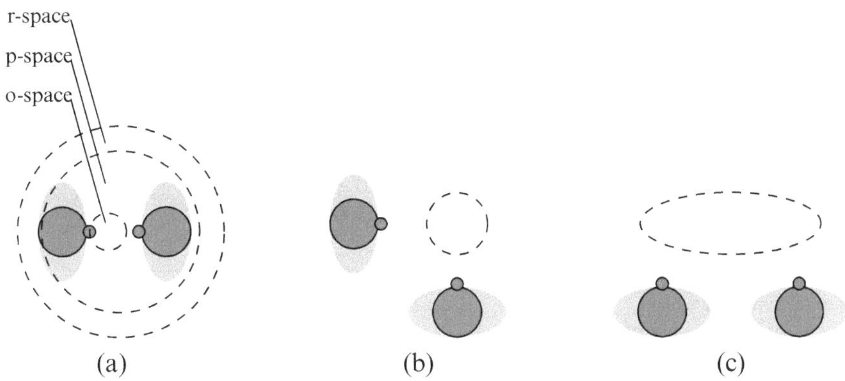

Fig. 7.1 Several types of F-Formations: (**a**) vis-a-vis, (**b**) L-shape (**c**) side-by-side

produced by activities of interacting agents. According to Kendon, three regions can be distinguished within such an F-Formation (see Fig. 7.1): The participating agents are located in the *agent region* (p-space). There is a common *transactional region* (o-space), in which most of the activity takes place, i.e., the region into which the agents look, speak, or where they handle objects. In addition, the *buffer region* (r-space) separates the activity from the rest of the environment. Each of those regions carries a social meaning to both the participants of the activities and to non-participants (Ciolek & Kendon, 1980). E.g., non-participants avoid crossing the transactional region or entering the buffer region. Agents that intend to participate in the ongoing activity signal their intention in the buffer region before they enter the agent region and become part of the activity. Depending on the relative orientation of the participants, Kendon (1990, p. 213) distinguishes three kinds of F-Formations called vis-a-vis, L-shape, and side-by-side (see Fig. 7.1).

While Kendon focuses on the spatial structure of interactions between humans, we are generally concerned with activities of humans or robots involving some environmental entity, which affords the activity. However, in Kendon's analysis, environmental entities affording an activity are not considered.

Therefore we take Kendon's analysis as the basis for our spatial model of activities and affordances but have to generalize it in two respects. On the one hand, we include in the model activities of individual agents in addition to activities of interacting agents, on the other hand, we allow to distinguish environmental objects as passive participants from agents as active participants of an activity. Furthermore, we assume that buffer regions are relevant for cases of complex activities such as interactions of agents only. Activities involving only one agent and an environmental entity do not necessarily induce buffer regions.

Our model of activity spaces provides three basic regions: agent region, transactional region, and the *patient region*, which is the region occupied by environmental entities that are passive participants in the activity (cf. Figs. 7.2 and 7.3). Further-

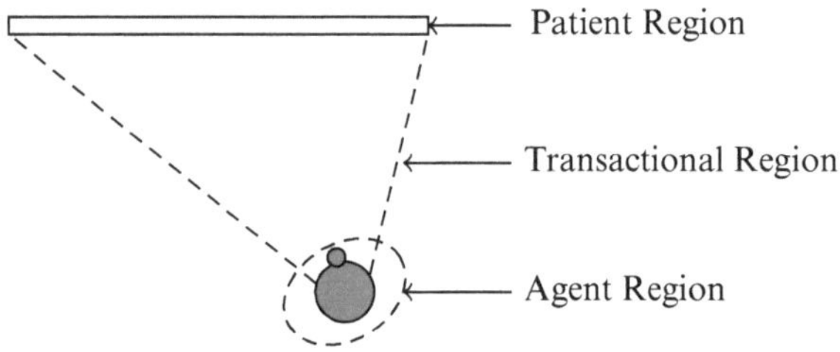

Fig. 7.2 Scheme of the activity space of an activity of viewing a picture

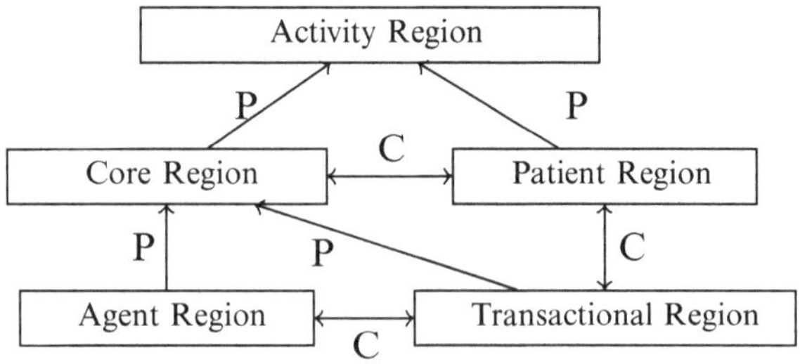

Fig. 7.3 The social regions of an activity space. The specification makes use of the relations parthood (*P*) and connection (*C*). If a region is part of another region, then these regions are also connected

more, the union of the agent region and the transactional region is called *core region* and the union of the core region and the patient region is called *activity region*.

The term *activity space* is used to refer to a structure that identifies the five named regions. The regions are also called *activity-space regions* or *social regions* of the acitivity space as they carry different social meanings for participants and non-participants of the activity.

The extension and geometry of the social regions within an activity space depends on the type of activity and spatial properties of the participants. However, the transactional region of an activity space is connected to both the agent region and the patient region. But depending on the activity and the granularity of modelling, the transactional region might also be part of the agent region (e.g., in modelling reading a book by a human, it might not be worth distinguishing a transactional region from the agent region).

The agent region of an ongoing activity is occupied by the agent. Therefore other agents should avoid to use it at the same time as the agent region of their activities or pass through it. Also placing oneself in the transactional region or even passing through a transactional region yields a disturbance of the other's activity and should be avoided. However, passing through a transactional region is less critical than passing through an agent region or standing in the transactional region for a longer time. To avoid disturbing ongoing activities activity spaces of ongoing and planned (potential) activities have to be tested for conflicts. Affordance spaces provide a means to model the spatial structure of potential activities.

7.3.2 General Structure of Affordance Spaces

In analogy to the notion of affordances as potential activities, affordance spaces are potential activity spaces (Lindner & Eschenbach, 2013; Lindner, 2015). Environmental entities function as source or host of an affordance. Such an entity providing an affordance is called *affordant* in the following. If an activity realizes an affordance, then the affordant is a (passive) participant in the activity and the activity produces an activity space. The spatial structure produced by an affordance corresponds to the structure of the activity (see Fig. 7.4): An affordance space provides a *potential agent region*, a *potential transactional region*, and an *affordant region*. We call the union of the potential agent region and the potential transactional region the *potential core region*. The union of the potential core region and the affordant region is called the *affordance region*. The five regions provided by an affordance are also called *affordance-space regions* or *(social) regions* of the affordance space.

While an activity is taking place, the affordance space used by the activity turns into an activity space (see Fig. 7.5). We define an activity space as follows: An affordance space *sp* related via an affordance to an activity type Φ is an activity space at some time point *t* if and only if at *t* there is an activity of type Φ taking place at *sp*, whereby the affordant region of the affordance space turns into a patient region, the potential agent region turns into an agent region, and the potential transactional region turns into a transactional region. Hence, the potential core region of the used affordance space turns into the core region of the activity space and the affordance region turns into the activity region. For instance, while viewing

Fig. 7.4 The (social) regions of an affordance space

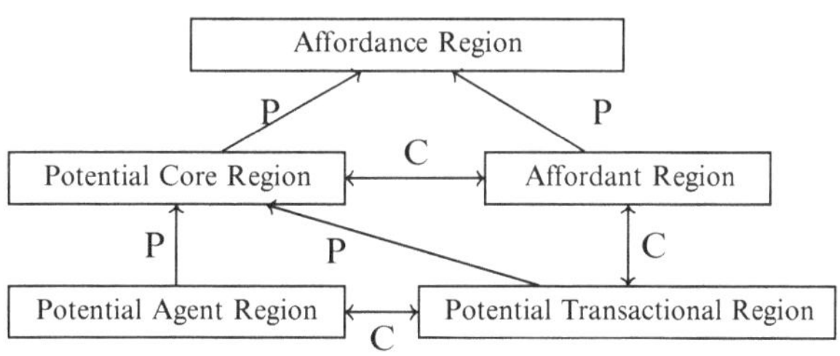

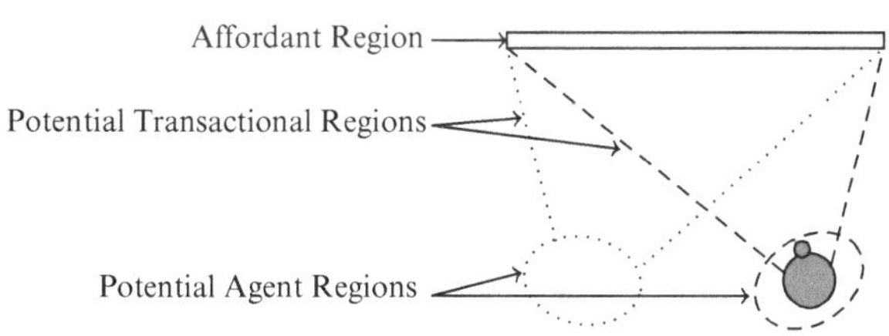

Fig. 7.5 Two affordance spaces produced by the picture's affordance of being viewable. An unused affordance space (*dotted lines*) and an affordance space used by an activity (*dashed lines*), i.e. an activity space

a painting, the affordance space produced by the viewing affordance turns into an activity space produced by the actual viewing.

Some activities such as viewing a painting do not use the transactional region physically, that is, while the activity takes place, the transactional region is kept free. This is not true for activities such as watering a flower, where the transactional region is filled by the watering can and water. We say that a potential transactional region is *physically used* if and only if the instances of the activity type the affordance space is related to use the transactional region physically.

7.3.3 Affordances and Affordance Spaces

Affordance spaces are *produced by* affordances and represent generic spatial constraints for the afforded activities. Figure 7.6 summarizes the interrelations between the concepts we use to describe affordances and affordance spaces. A formal axiomatization of the conceptual model can be found in (Lindner, 2015).

Affordances are properties (dispositions) that *inhere in* affordants. Thus, we adopt the view of Turvey (1992) according to which affordances are properties of environmental entities such as objects and living beings. Activities *realize* affordances but affordances exist independently of the activity taking place. Therefore, we model affordances as primarily related to *activity types enabled* by affordances. Activities are *performed by* agents and can have an affordant as *participant*. If a robot grasps a bottle, then the robot is agent of and the bottle is participant in the grasping activity. An agent of an activity requires certain *abilities used* in the activity fitting to the affordances of the affordant for success. These abilities include spatial

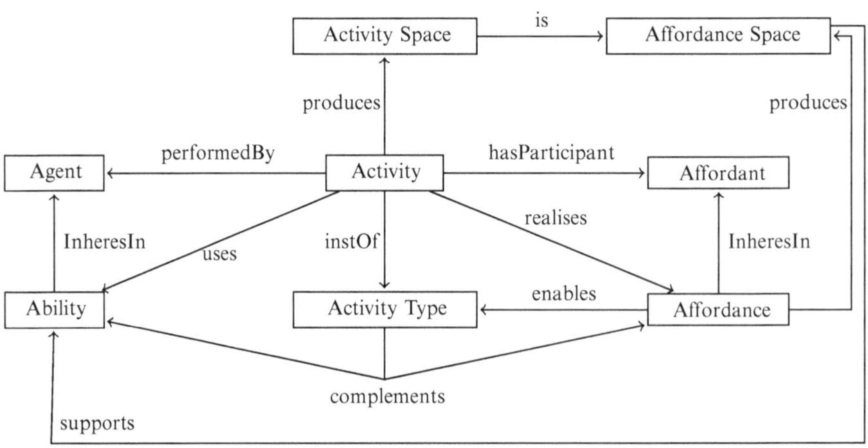

Fig. 7.6 The interrelations between concepts in the theory of affordances and affordance spaces

requirements of the agent. Thus, if a robot (agent) grasps a bottle, the grasping (activity) uses the robot's manipulation abilities. If the agent's abilities fit to the affordant's dispositions for carrying out the enabled activity type, then we say that the abilities *complement* the affordances relative to the activity type. Hence, if the robot's abilities complement the bottle's affordance with respect to grasping, then we can say that the bottle affords grasping to the robot. When the robot grasps the bottle, then the activity of grasping uses the robot's ability and realizes the object's affordance. Again, the idea of abilities and affordances being complementary is in line with Turvey's affordance ontology (Turvey, 1992).

The structure of affordance spaces derives from the spatial needs of activities as discussed in Sect. 7.3.2. The shape and size of the regions provided by an affordance space depend on the type of the afforded activity, the spatial structure of the affordant, the (spatial) abilities of the participants, and the surrounding. Affordants determine the reference frame for affordance-space regions and activity-space regions. Thus, if an affordant moves, the corresponding regions move as well. We also say in such cases that the affordance space is *attached to* the affordant.

Affordance spaces reflect the spatial constraints deriving from complementing affordances and abilities. The affordances of the affordant and abilities of (potential) agents determine the shape and size of the regions of activity spaces and of affordance spaces produced by the activities and affordances, respectively. For instance, a robot with a short manipulator needs to move closer to the bottle to grasp it than a robot with a long manipulator (these properties belong to the realm of abilities). On the other hand, the robot with the long manipulator might need to keep a certain distance to the bottle to be able to move the manipulator as needed. Thus, the potential agent region produced by the grasping affordance varies in size and distance to the affordant with the abilities of the agent (e.g., it might be more distant and bigger for the long-armed robot). To be able to clearly map between different abilities of agents and fitting geometries of affordance-space regions, an affordance can produce different affordance spaces that *support* different abilities.

We consider affordance-space regions to be vulnerable to spatial behavior comparable to activity-space regions. In particular, by selecting a certain position for carrying out an activity (e.g., for recharging) potential agent regions or potential transactional regions related to some other kind of activity can be blocked, and thus, action possibilities for agents become deactivated. However, the number of affordance spaces included in an environment might be rather large, as, for example, every object and every piece of wall affords viewing. Therefore, we need the possibility to distinguish affordances spaces that are socially relevant (we call them *social affordance spaces*) and evaluate potential conflicts derived from blocking them. In Sect. 7.5 we argue that social affordance spaces are affordance spaces to which socio-spatial reasons against blockages apply. Socio-spatial reasons derive from rules that take expected activities of other agents in the environment into account. But before going into the details of this model, we give some examples to show that afforcance-space knowledge can be used in the context of different types of positioning tasks.

7.4 Affordance-Space Awareness

Knowledge about affordances and affordance spaces can be exploited on different levels of planning and acting. First, if an agent acts on an object, knowledge about the affordances provided by the object helps to act successfully. For example, to successfully grasp a bottle, a robot needs to move into an appropriate area and approach the bottle with its manipulator from a specific direction. Thus, affordance knowledge can be employed on the functional level. This use of affordances might not require the ability to take the perspective of other agents on the affordances into account. As mentioned in Sect. 7.2, different approaches in robotics address the acquisition of functional affordance knowledge and its spatial aspects.

Robots are artifacts themselves that provide affordances to humans. When a robot moves, the affordance spaces it provides move as well. Similarly, we can also ascribe affordance spaces to affordances provided by humans regarding interaction. Thus, the second use of knowledge about affordances concerns the interaction of two or more agents based on interaction affordance spaces. Successful interaction often requires spatial coordination, such that the agents have to move to affordance spaces provided by other participants.

An example concerning affordances of interaction between a human and a robot is depicted in Fig. 7.7. The robot provides an information screen that can be viewed by humans with intact vision producing an affordance space sp_v. The human affords being talked to by the robot, and therefore, corresponding affordance spaces are available.

In situations as in Fig. 7.7a interaction is not possible. To interact with the robot, the human has to be located in the agent region of sp_v, i.e., to be sufficiently close to see the display. Moreover, the human has to be oriented towards the display, i.e., according to the potential transactional region of sp_v. Similarly, to talk to the human the robot has to be in the agent region of a fitting affordance space.

When both agents are located in respective agent regions as in Fig. 7.7b, bilateral interaction can take place. During interaction, the human viewing the screen occupies the agent region of sp_v, and the robot talking to the human occupies the agent region of an affordance space of the human.

Initially, we have started out by adopting Kendon's analysis of F-Formation and fit it to the case of single agents acting upon affordances. Now we see that Kendon's original F-Formations emerge from several activity spaces, viz., they are constituted by those activity spaces used by agents that interact with one other (Fig. 7.7c).

Third, affordance knowledge can be employed on a social level to avoid disturbing other agents in their planned activities. In this case, potential activities of other agents and their spatial requirements have to be taken into account.

In Fig. 7.8, there is a doorway and a robot. Among others, the doorway affords moving through to humans. The robot's task is to provide information to humans and therefore it has a monitor mounted on its body that affords viewing to humans. Consequently, there are two affordances constituting two affordance spaces.

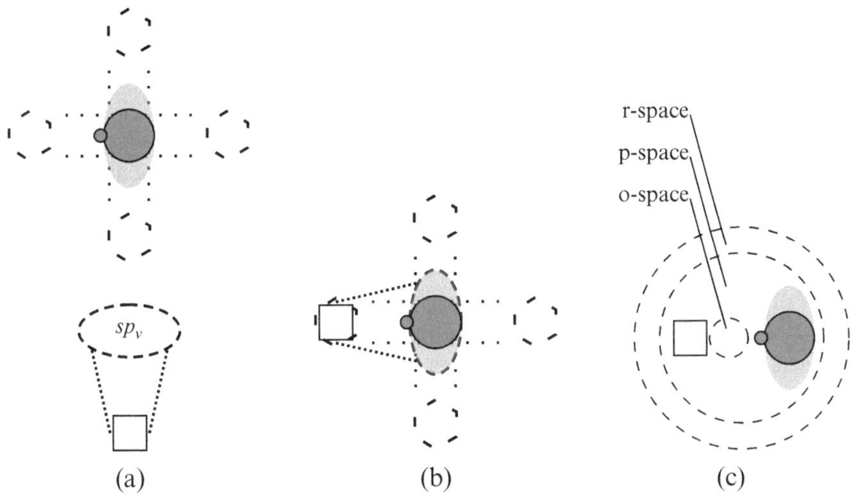

Fig. 7.7 Simplified illustration of four affordance spaces produced by the talking affordance inherent in a human and one affordance space produced by the viewing affordance inherent in a robot (*square*) with potential agent regions (*dashed*) and potential transactional regions (*dotted*). In (**a**) interaction is hindered by the separation of the affordance spaces. In (**b**), the robot occupies the potential agent region of a fitting affordance space of the talking affordance of the human, and the human occupies the potential agent region of a fitting affordance space of the viewing affordance of the robot, the viewing area aligned with the potential transactional region as necessary for interaction. (**c**) shows the F-Formation constituted by the interaction of the human with the robot

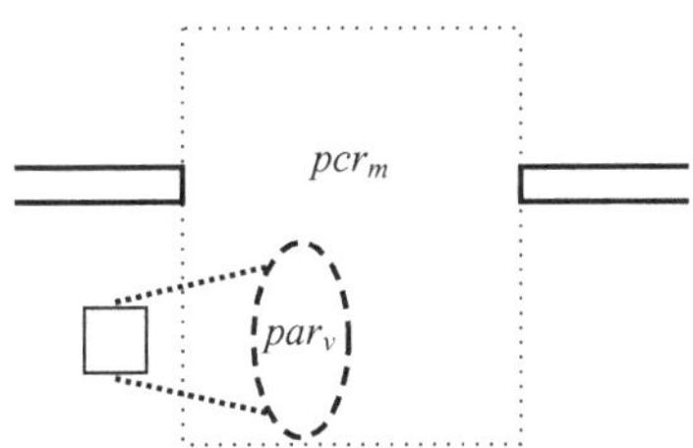

Fig. 7.8 A potential conflict due to the spatial relation between two affordance spaces. The potential core region (pcr_m) of the movement affordance provided by the doorway intersects the potential agent region (par_v) of the viewing affordance provided by the robot (hexagon)

In the spatial setting sketched in Fig. 7.8 the robot is positioned next to the door to provide information to people entering the room. It does not block the doorway and therefore people who do not need the information can just pass by. However, this arrangement might easily provoke conflicts, as the potential core regions of the two affordance spaces partially coincide. If a human would start to interact with the robot, the human would be located in the potential agent region (par_v) of the viewing affordance. Thus, the human would also be located within the potential core region (pcr_m) of the movement affordance, i.e., partially blocking the region that is needed for other humans to move through the doorway. Thus, a socially aware robot needs to be aware of affordance spaces produced by itself and by other entities in its vicinity to be able to evaluate spatial configurations with respect to their social acceptability.

The final example in this section shall demonstrate that the spatial requirements associated to an affordance also depends on the abilities of agents that act upon the affordances. Service robots that deliver fresh supplies (e.g., food trays, sheets) need to be able to decide about where to leave the supplies in case there is no human agent available to directly take care of them. For humans, it is obvious that the center of a corridor, a door, or the area in front of a light switch are not the regions to choose to deposit such supplies, even though there might not be anyone present who acts in those spaces.[2] Both corridors and doorways are environmental entities that afford movement. The area to be kept free depends on the expected traffic, the size of the agents, the objects they transport, and on abilities of the potential agents. Correspondingly, people prefer to deposit goods along walls and keep the center of a corridor free for traveling. However, if hand railings are mounted on a corridor wall, the corridor produces an affordance space for people who prefer to move supported or guided by such railings. In this case, placing objects along the wall blocks this affordance space and thereby deactivates the affordance for people using railings. When different types of agents with different abilities populate an environment, different affordance spaces might be associated with one affordance. Thus, the presence of a railing on one wall of the corridor does not justify the assumption that supplies can be deposited in the center of the corridor, as other agents might want to move there. To be able to identify the affordance spaces that are relevant on the social level within an environment shared by different types of agents, a socially aware robot needs knowledge about which activities are to be expected by its co-inhabitants of the environment.

The examples in this section show that knowledge about affordances and affordance spaces can be employed in a variety of action and planning tasks. Correspondingly, several approaches in robotics are concerned with modelling the spatial constraints of (inter-)actions (e.g. Yamaoka et al., 2008; Sisbot et al., 2010). Although the focus of such work might be directed to a specific task such as the robot's task to position itself to provide information or to hand objects to some person, the same knowledge can be employed to avoid to park in such regions when interaction with a human is not intended. However, the multiple use of knowledge about activities and affordances requires linking the specific action models to a general framework for modelling activities, affordances and the social spaces they produce.

[2]Regions that need to be free of obstacles for safety reasons can be marked in maps. However, more general mechanisms that model potential activities and their spatial requirements might be called for when service robots are brought into environments that are not completely mapped beforehand and that contain mobile artifacts providing affordances.

7.5 Reason-Driven Activity-Placement Planning

In the following, we focus on solving the *social activity-placement problem*, which is defined as follows: *Given an activity type ϕ and a potential agent α, a placement π should be determined, such that α can successfully ϕ at π, and π is among the most socially adequate placements for α to ϕ* (Lindner & Eschenbach, 2013). The activity-placement problem is a generalization of the robot placement problem (e.g. Stulp, Fedrizzi, Mösenlechner & Beetz, 2012; Zacharias, Borst, Beetz, & Hirzinger, 2008). The potential agent can be the robot itself, but it can also be some other agent for whom the robot has to find a placement.

To determine a most socially adequate placement for an activity, a social robot needs to identify locations where its own and others' activities can be placed (functional level), identify potential conflicts related to these placements (functional level), find the socio-spatial reasons that speak for or against the derived candidate placements relative to the identified potential conflicts, and determine the preference order of the placement candidates resulting from the related reasons (social level). In the first subsection, we will address the functional level, in the second subsection, we will address the social level on this basis.

7.5.1 Candidate Affordance Spaces and Conflicts

To establish the foundation for an affordance-based solution to the social activity-placement problem, we define candidate affordance spaces and conflicts between affordance spaces. An affordance space *sp* is a *candidate affordance space* for activities of the type ϕ as performed by an agent α if and only if the affordance that produces *sp* is complementary to an ability of α with respect to ϕ and *sp* supports the spatial ability of α. The available candidate affordance spaces in a given situation thus constitute the space of functionally possible placements for an activity of type ϕ performed by agent α.

With respect to the social level, we aim to implement a principle of social activity placement which says "You should not block others' possiblities for action." Therefore, we have to settle what it means to block a possiblity for action. To this end, we introduce definitions of types of conflicts between affordance spaces (cf. Lindner, 2015).

First, we observe that affordance spaces cannot be used by different activities in parallel if their potential agent regions overlap, because otherwise the agents of the two activities would physically collide. Thus, affordance spaces are in *conflict of type AA* if and only if their potential agent regions overlap. We observe that every affordance space is in *AA*-conflict with itself. This fits the demand that the conflict relation models which affordance spaces cannot be used by different activities (with different agents) at the same time.

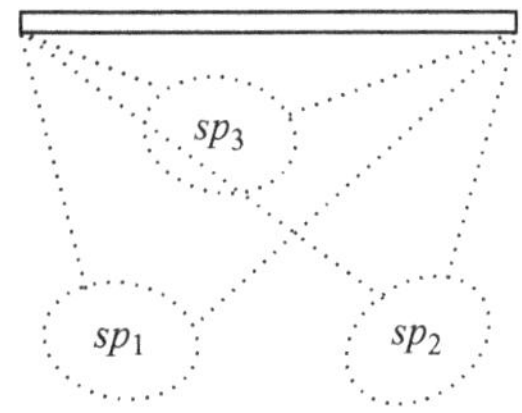

Fig. 7.9 Affordance spaces of the painting in conflict to each other

It is not necessarily the case that two affordance spaces are in conflict if their potential transactional regions overlap. For instance, the affordance spaces sp_1 and sp_2 in Fig. 7.9 can be used in parallel by two different visitors viewing the painting. However, if activities using the affordance spaced make conflicting physical use of the transactional regions (for instance, activities of painting), then the affordance spaces should be classified as being in conflict. Affordance space *sp conflicts with type TT* with affordance space sp' if and only if the potential transactional regions of sp and sp' overlap and the physical use associated with the potential transactional region of sp conflicts with the physical use associated with the potential transactional region of sp'. Note that while *AA*-conflict is a symmetric relation, this is not true for *TT*-type conflicts.

Furthermore, in Fig. 7.9 the affordance space sp_3 cannot be used in parallel without the agent of the activity placed in sp_3 standing in the field of view of the agent placed in sp_1 or sp_2. This is because the potential agent region of sp_3 overlaps the potential transactional regions of sp_1 and sp_2. This type of conflict is of the *AT*-type. Affordance space *sp conflicts with type AT* with affordance space sp' if and only if the potential agent region of sp overlaps the potential transactional region of sp' and the presence of an agent in the potential agent region of sp conflicts with the physical use associated with the potential transactional region of sp'. Correspondingly, affordance space *sp conflicts with type TA* with affordance space sp' if and only if the potential transactional region of sp overlaps the potential agent region of sp' and the physical use associated with the potential transactional region of sp conflicts with the presence of an agent in the potential agent region of sp'.

We say that two (not necessarily distinct) affordance spaces are in *conflict* if and only if they stand in one of the discussed conflict relations. Sometimes, it turns out to be more convenient to subsume *AA*-conflicts and *AT*-conflicts under the concept of an *AC*-conflict, which is a conflict due to the overlap of a potential agent region with a potential core region: If and only if two affordance spaces stand in conflict of type *AA* or of type *AT*, they stand in conflict of type *AC*.

7.5.2 Socio-Spatial Reasons

So far, the model contains candidate affordance spaces in which activities can be placed as well as conflict relations between affordance spaces. However, this information is not yet sufficient to solve the social activity-placement problem for at least two reasons: First, not every conflict yields a real problem. Think of all the affordance spaces around us no one might ever (or rarely) use, e.g., the affordance spaces attached to the wall which can be used to hang up a picture. Usually, it is not expectable that someone uses such affordance spaces and therefore we block such affordance spaces all the time without being accused of inconsiderate behavior. Second, often there is no way to avoid conflicts. In such dilemma situations, to pick the socially most adequate candidate affordance space for an activity, one needs a means to rank the available candidates based on the identified conflicts.

To provide normative meaning to the relevant affordance spaces and to be able to rank candidate affordance spaces, the concept of a socio-spatial reason is introduced and put into action using a decision rule.

Our reason-driven view (cf., Lindner, 2015) is inspired by contemporary work in practical philosophy. Particularly, the moral philosopher Joseph Raz characterizes reasons as "facts that have normative bearing or force" (Raz, 2011, p. 23). Furthermore he claims that "[Reasons] are called 'reasons' because they can serve as stepping-stones in reasoning about what to believe or what to do. Deliberating from the reasons that apply to us we become aware of the attractions and drawbacks of options" (ibid).

According to this characterization, we introduce *socio-spatial reasons* as a special sort of reasons. They are normative facts like "one should not block this affordance". As such they are connected to particular affordances via a relation called *source-of*. For instance, in Fig. 7.8 (p. 147), the affordance inherent to the doorway is source of a reason "one should not block the affordance to walk through this door". Socio-spatial reasons *apply to* agents placing activities at affordance spaces. In the case depicted by Fig. 7.8, the reason applies to placing viewing activities in the affordance space attached to the robot.

In addition to being sourced by entities such as affordances, socio-spatial reasons are modeled as having normative force. Normative force is assumed to take values in a two-dimensional space spanned by polarity (pro or con) and strength (e.g., strong, medium, and weak). For instance, the fact "you should not block the affordance to walk through this door" is a con-reason. The strength might depend on different factors like how many other doorways there are to enter this area and how many humans are expected to move into and out of this area. On the other hand, affordances are sources of pro-reasons like "one should behave as expected". For instance, if a particular robot usually recharges its batteries at a particular outlet such that the others expect the robot to use it if needed, the robot should keep using this particular outlet, that is, the affordance space attached to it.

Having said that, the activity-placement problem, viz. the problem to place an activity of type ϕ as performed by agent α, can be solved by a three-step procedure:

Table 7.1 Example bipolar decision case: Candidate affordance spaces sp_1, sp_2, sp_3 and reasons $\rho_{1\text{-}8}$ having polarity (pro/con) and strength (strong, medium, weak)

	sp_1		sp_2		sp_3	
	Pro	Con	Pro	Con	Pro	Con
Strong					ρ_6	
Medium	ρ_1		ρ_3			
Weak		ρ_2		ρ_4, ρ_5	ρ_7	ρ_8

First, the available candidate affordance spaces that can be used for ϕ-activities as performed by α are determined. Second, for each of these affordance spaces the socio-spatial reasons that apply to them are identified. Third, a decision rule ranks the candidate affordance spaces based on aggregating the reasons and proposes a best candidate based on this ranking.

Among the decision rules available, we choose the Levelwise Tallying rule (Dubois, Fargier, & Bonnefon, 2008). Bonnefon, Dubois, Fargier, and Leblois (2008) show Levelwise Tallyingto predict many human choices in a psychological experiment.

Our adaption of Levelwise Tallying determines for each pair of candidate goal affordance spaces sp_1 and sp_2 whether the usage of sp_1 is to be preferred over the usage of sp_2. For both candidates, the number of con-reasons of the highest strength are subtracted from the number of the pro-reasons of highest strength. The candidate with the highest score wins. If the scores are equal, then the procedure is repeated on the next lower level of strength. If neither of the candidates is preferred to the other, they are ranked equivalent regarding preference.

An example is given in Table 7.1: A decision has to be made between using affordance space sp_1, sp_2, or sp_3. Since sp_3 is the only one with a pro-reason at the highest level and there is no con-reason at the same level, it wins the comparison with sp_1 and sp_2. Affordance space sp_1 is preferred to sp_2 as the reasons at the two highest levels are in balance and sp_2 has more con-reasons than sp_1 on the lowest level. Thus, the candidates are ordered as $sp_3 > sp_1 > sp_2$. After ranking the candidate goal affordance spaces, one of the best candidates can be selected (in this example, sp_3 will be selected).

7.6 Case Studies

This section demonstrates by two case studies how the social activity-placement problem can be solved by employing the concepts and relations so far defined. The case study in Sect. 7.6.1 is about a human employing socio-spatial reasons to make a choice between two available candidate affordance spaces. Section 7.6.2 exemplifies how an affordant can use knowledge about affordance spaces and socio-spatial reasons to place itself attentively.

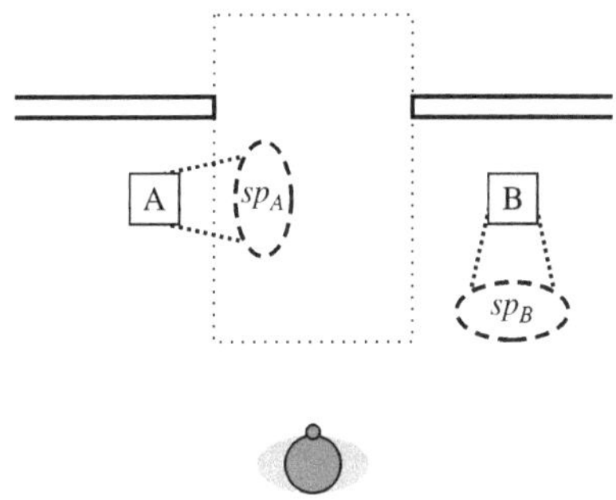

Fig. 7.10 Activity Placement: The human has the choice to either place its activity at the affordance space sp_A of affordant A or at the affordance space sp_B of affordant B

7.6.1 An Activity-Placement Scenario

In Fig. 7.10, there are a doorway, two terminals A and B, and a human H. The doorway inheres an affordance complementary to abilities of mobile actants with respect to the activity type of walking through the door. This doorway's affordance produces an affordance space depicted as a dotted rectangular in Fig. 7.10. Besides the doorway, there are two more affordants A and B. For the sake of this example, we assume A and B to be ATMs. Each of them inheres an affordance which is complementary to the ability of the human H to draw money at an ATM. Each of these affordances produces an affordance space at which the human can place its activities of type *drawing money*. The affordance space attached to ATM A is represented by sp_A and the one attached to ATM B is represented by sp_B.

Let us now assume that the human H wants to place an activity of the type *drawing money* socially, assuming that there are other humans around that want to pass through the doorway. That is, the human has to solve an instance of the social activity-placement problem: Given the activity type *drawing money* and the potential agent H, a placement π should be determined, such that H can successfully do *drawing money* at π, and π is among the most socially adequate placements for H to do *drawing money* (compare to the definition of the general social activity-placement problem on page 149).

The first step of the procedure is to determine the candidate affordance spaces for *drawing money* activities performed by H. These are the affordance spaces of A (sp_A) and of B (sp_B). Subsequently, the socio-spatial reasons are identified that apply to using either of these candidate affordance spaces. Based on the assumption that someone else will likely use the affordance space of the doorway, the socio-spatial reason of negative polarity provided by the affordance of the doorway applies to H using sp_A for drawing money as the sufficient conditions are fulfilled in this situation: The candidate affordance space sp_A is in conflict (of type AC) with an affordance space that someone else might want to use.

Hence, in this situation, the socio-spatial reason "you should not block the affordance of walking through the door" is a reason against H using sp_A for drawing money, but not against H using sp_B for drawing money. Consequently, the Levelwise Tallyingrule outputs the order $sp_B > sp_A$.

Acting in a socially aware manner, the human H will now place its activity at sp_B.

7.6.2 An Affordant-Placement Scenario

The task of affordant-placement shows that a social robot that knows about the affordance spaces attached to one's own body can attentively position itself such that someone who wants to act upon one of the robot's affordances can do so without blocking the affordance spaces of others.

A corresponding situation is depicted in Fig. 7.11. Unlike the first example, A is the only ATM available and A can move.[3] As in the previous example there is a doorway affording walking through the door. The ATM A provides the affordance to withdraw money. This affordance produces the affordance space sp_1. In addition, A affords to be reloaded. Thus, if the fill status of A drops below a certain level, A can alert people who have the permission to open and refill A. This affordance produces affordance space sp_2.

In the situation depicted in Fig. 7.11, the affordance space sp_1 is in conflict with the affordance space of the doorway, and affordance space sp_2 cannot be used because it overlaps with the wall. We assume that the robot can perform 90 degrees clockwise rotations and forward translations in half-meter steps. At each pose so reachable from the initial pose, the robotic ATM can identify the conflicts between affordance spaces using its knowledge about affordance spaces and conflicts. For instance, just using the knowledge about conflicts of affordance spaces, the ATM might decide to move to the bottom left corner, where none of the affordance spaces attached to A conflicts with the affordance space of the doorway. If none of the affordance spaces conflicts with the affordance space of the doorway, the socio-spatial reason not to block the doorway does not apply to any of them.

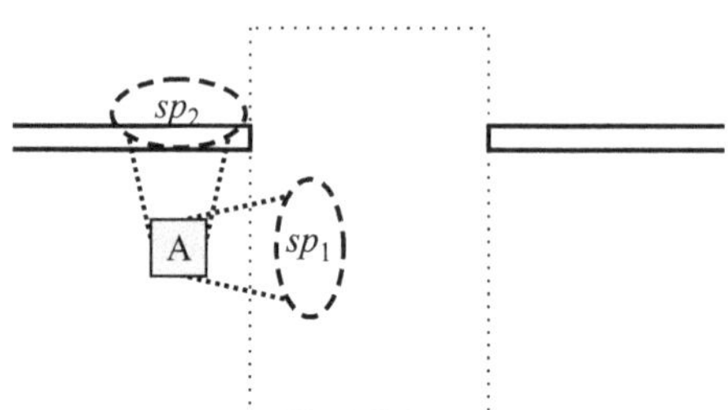

Fig. 7.11 Affordant placement: In the current situation, the affordance space sp_{A_1} attached to the robot cannot be used by a human without blocking the door. The affordance space sp_{A_2} cannot be used because it overlaps with the wall

[3]There already exists a robotic ATM prototype. Such machines might be deployed in casinos as pointed out in an online article that can be found on the internet: http://www.atmmarketplace.com/blogs/cash-at-your-service/

Figure 7.12 illustrates how the robot can systematically find a pose where none of its affordance spaces are in conflict with the affordance space of the doorway. Starting from the initial situation (S1) the robot successively simulates rotations and forward translations. After two virtual rotations, a state (S4) is found which fits the demand that both affordance spaces sp_1 and sp_2 can be used without conflicts.

However, one might also be interested in minimal sequences of actions that get the robotic ATM from the initial pose to a pose which is optimal with respect to the current expectations about the use of the affordance spaces sp_1 and sp_2. Assume that it is expectable that the human H uses sp_1 for withdrawing money. In this case, the search for a better pose can already be terminated after one rotation, because the resulting state (S2) has the property that there is no socio-spatial reason speaking against H using sp_1 for withdrawing money. Note that this state is suboptimal with respect to using affordance space sp_2.

A different situation occurs when the fill status of the ATM A drops and A sends an alert to someone who is authorized to refill the ATM (let us call this person $H2$). In this case, the ATM being in the initial situation realizes that there is no candidate affordance space usable that could be used by $H2$ refilling it. Thus, the robot invokes the search process and successively evaluates the resulting states with respect to the socio-spatial reasons that apply to $H2$ using sp_2 for refilling the ATM. Consequently, the search process in Fig. 7.12 finds a socially adequate pose after two rotations (S4). This is the same state found in the search for a state with no conflicts between the affordance spaces of the ATM and the affordance space of the robot. If we allow the robot to also perform counterclockwise rotations or sideway translations, the search process will find a pose which guarantees $H2$ to use sp_2 in a socially adequate manner after either one of these two actions.

As is once more shown by this example case for the affordant-placement task, the mere location of an affordant does not suffice to evaluate the social acceptability. Particularly, it is not sufficient to only check if A is in the doorway, because there are poses of A that lead to conflicts although A itself is not blocking the doorway at all. Only when the knowledge about the affordance spaces is represented, and thus the whole space that will be needed by activities, activities can be placed such that a potential user will not be accused of socially inadequate behavior.

7.7 Discussion and Outlook

Affordances are the possibilities for activities provided by environments to agents. It is well-established that activities are spatially extended. Hence, activities have spatial requirements that depend on the type of the activity as well as on the participants' spatial properties (dispositions and abilities). Consequently, we observe that affordances structure space according to the spatial requirements of the afforded activities taking the abilities of potential agents into account.

Defining affordance spaces as depending on affordances and on abilities of potential agents, our proposal allows to model spatial aspects of affordances in

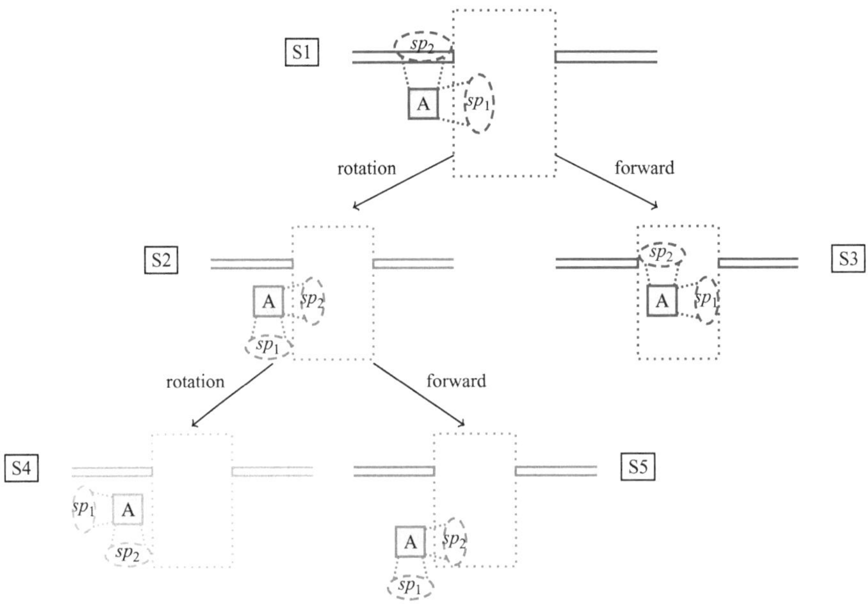

Fig. 7.12 Departing from the initial situation (*S1*) the space of possible poses is explored by virtally applying rotation and forward translation to the robot's position. Each resulting state is evaluated. *S1* and *S3*: Both of the robot's affordance spaces sp_1 and sp_2 cannot be used without conflict. *S2* and *S5*: At least one of the affordance spaces can be used without blocking the door. *S4*: Both of the robot's affordance spaces can be used without any reasons speaking against it

a flexible manner. Taking affordances and their affordance spaces into account, different perspectives on the structure and use of space can be derived.

The general framework for modelling affordances, activities, and their spatial requirement proposed in this article provides a basis for using knowledge about affordances and activities in a diversity of reasoning scenarios. The knowledge about the potential agent region of a viewing affordance provided by a screen mounted at one side of a robot can be used by the robot to reason both about where to move and how to orient to show certain information to a human, or to reason about where to wait for humans to come up when they seek information. Such reasoning would combine specific models regarding the size and shape of agent region and transactional region for specific activities with general rules regarding possible configurations of affordance spaces and activity spaces.

The acquisition of data for building models for optimal positioning of robots in specific activities is a complex and time-consuming task (see, e.g. Stulp et al., 2012; Zacharias et al., 2008; Yamaoka et al., 2008). However, when such a model is related to the general framework for modeling affordances and affordance spaces presented in this paper, it can be employed in a variety of situations.

One general rule can be derived from the observation that potential agent regions of affordance spaces are occupied by agents when they act upon the affordance.

Therefore, blocking (access to) potential agent regions and overlap between potential agent regions of different affordance spaces are conflict prone and should be avoided when one expects the usage of the affordance spaces by co-inhabitants of the environment. On the other hand, motion through (unused) affordance spaces is much less problematic than motion through activity spaces. Further principles of socially appropriate spatial behavior are formalized in (Lindner, 2015).

The reason-driven approach to modeling social affordance spaces gives a basis for communicating why a certain choice has been taken ("I recharge here as my battery is very low and recharging at the other recharging stations close by would disturb John watching TV or block the entry to the kitchen"), generating an apology ("I am sorry to block the entry to the library"), or doing both ("I am sorry to block the entry to the library but I have to recharge and all other stations are occupied by other robots").

Finally, from the perspective of robot behavior programming, it is noticable that the socially-aware robots presented here have the capability to adapt their spatial behavior autonomously. There is no need to program the robots, for instance, to move to one pose at eight p.m. and stay there until ten p.m. and then to move to another pose etc. Under the affordance-based and reason-driven perspective, all decisions rest upon the knowledge declared in the robot's knowledge base.

References

Bonnefon, J.-F., Dubois, D., Fargier, H., & Leblois, S. (2008). Qualitative heuristics for balancing the pros and cons. *Theory and Decision, 65*(1), 71–95.

Chemero, A. (2003). An outline of a theory of affordances. *Ecological Psychology, 15*(2), 181–195.

Ciolek, T. M., & Kendon, A. (1980). Environment and the spatial arrangement of conversational encounters. *Sociological Inquiry, 50*(3–4), 237–271.

Dohn, N. B. (2009). Affordances revisited: Articulating a Merleau-Pontian view. *International Journal of Computer-Supported Collaborative Learning, 4*(2), 151–170.

Dubois, D., Fargier, H., & Bonnefon, J.-F. (2008). On the qualitative comparison of decisions having positive and negative features. *Journal of Artificial Intelligence Research, 32*(1), 385–417.

Gibson, J. J. (1977). *The Theory of Affordances*. In R. E. Shaw & J. Brandsford (Eds.), Perceiving, acting, and knowing: Towards an ecological psychology (pp. 67–82). Hillsdale, NJ: Erlbaum.

Kendon, A. (1990). *Conducting interaction. Patterns of behavior in focused encounters*. Cambridge/New York: Cambridge University Press.

Krüger, N., Geib, C., Piater, J., Petrick, R., Steedman, M., Wörgötter, F., ... Dillmann, R. (2011). Object-action complexes: Grounded abstractions of sensory-motor processes. *Robotics and Autonomous Systems, 59*(10), 740–757.

Kuhn, W (2007). An image-schematic account of spatial categories. In S. Winter, M. Duckham, L. Kulik, & B. Kuipers (Eds.), *Spatial information theory* (pp. 152–168). Berlin: Springer.

Lindner, F. (2015). *Soziale Roboter und soziale Räume: Eine Affordanz-basierte Konzeption zum rücksichtsvollen Handeln*. PhD Thesis, University of Hamburg.

Lindner, F., & Eschenbach, C. (2011). Towards a formalization of social spaces for socially aware robots. In M. Egenhofer, N. Giudice, R. Moratz, & M. Wor boys (Eds.), *Spatial information theory* (pp. 283–303). Berlin/Heidelberg: Springer.

Lindner, F., & Eschenbach, C. (2013). Affordance-based activity placement in human-robot shared environments. In G. Herrmann, M. J. Pearson, A. Lenz, P. Bremner, A. Spiers, & U. Leonards (Eds.), *Social robotics. 5th International Conference. ICSR 2013* (pp. 94–103). Cham: Springer.

Norman, D. A. (1999). Affordance, convention, and design. *Interactions, 6*(3), 38–43.

Raubal, M. (2001). Ontology and epistemology for agent-based wayfinding simulation. *International Journal of Geographical Information Science, 15*(7), 653–665.

Raubal, M., & Moratz, R. (2008). A functional model for affordance-based agents. In E. Rome, J. Hertzberg, & G. Dorffner (Eds.), *Towards affordance-based robot control* (pp. 91–105). Berlin/Heidelberg: Springer.

Raz, J. (2011). *From normativity to responsibility*. Oxford/New York: Oxford University Press.

Saffiotti, A., & Broxvall, M. (2008). Affordances in an ecology of physically embedded intelligent systems. In E. Rome, J. Hertzberg, & G. Dorffner (Eds.), *Towards affordance-based robot control* (pp. 106–121). Berlin/Heidelberg: Springer.

Şahin, E., Çakmak, M., Doğar, M. R., Uğur, E., & Üçoluk, G. (2007). To afford or not to afford: A new formalization of affordances toward affordance-based robot control. *Adaptive Behavior, 15*, 447–472.

Sisbot, E. A., Marin-Urias, L. F., Broquère, X., Sidobre, D., & Alami, R. (2010). Synthesizing robot motions adapted to human presence. *International Journal of Social Robotics, 2*(3), 329–343.

Stoffregen, T. A. (2003). Affordances as properties of the animal-environment system. *Ecological Psychology. 15*, 115–134.

Stoytchev, A. (2008). Learning the affordances of tools using a behavior grounded approach. In E. Rome, J. Hertzberg, & G. Dorffner (Eds.), *Towards affordance-based robot control* (pp. 140–158). Berlin/Heidelberg: Springer.

Stulp, F., Fedrizzi, A., Mösenlechner, L., & Beetz, M. (2012). Learning and reasoning with action-related places for robust mobile manipulation. *Journal of Artificial Intelligence Research, 43*, 1–42.

Turvey, M. (1992). Affordances and prospective control: An outline of the ontology *Ecological Psychology, 4*(3), 173–187.

Yamaoka, F., Kanda, T., Ishiguro, H., & Hagita, N. (2008). How close? A model of proximity control for information-presenting robots. In *Proceedings of the ACM/IEEE 3rd Annual Conference on Human-Robot Interaction* (pp. 137–144). Amsterdam, Netherlands.

Zacharias, F., Borst, C., Beetz, M., & Hirzinger, G. (2008). Positioning mobile manipulators to perform constrained linear trajectories. In *Proceedings of the IEEE/RSJ International Conference on Intelligent Robots and Systems* (pp. 2578–2584). Nice, France.

Chapter 8
Key Elements for Human-Robot Joint Action

Aurélie Clodic, Elisabeth Pacherie, Rachid Alami, and Raja Chatila

Abstract For more than a decade, the field of human-robot interaction has generated many valuable contributions of interest to the robotics community at large. The field is vast and addresses issues in perception, decision, action, communication and learning, as well as their integration. At the same time, research on human-human joint action has become a topic of intense research in cognitive psychology and philosophy, providing elements and even offering architecture hints to help our understanding of human-human joint action. In this paper, we analyse some findings from these disciplines and connect them to the human-robot joint action case. This work is a first step toward the development of a framework for human-robot interaction grounded in human-human interaction.

Keywords Action • Joint action • Architecture • Social robotics • Human-robot interaction

8.1 Introduction

For more than a decade, the field of human-robot interaction has generated many valuable contributions of interest to the robotics community at large. The field is vast, addressing perception (e.g., tactile or visual), decision (e.g., human-aware planning, supervision) and action (e.g., manipulation, navigation). At the same time,

---

A. Clodic (✉) • R. Alami
LAAS-CNRS, Université de Toulouse, CNRS, Toulouse, France
e-mail: aurelie.clodic@laas.fr; rachid.alami@laas.fr

E. Pacherie
Institut Jean Nicod, CNRS UMR 8129, Institut d'Etude de la Cognition, Ecole Normale Supérieure & PSL Research University, Paris, France
e-mail: elisabeth.pacherie@ens.fr

R. Chatila
Sorbonne Universités, UPMC, Univ Paris 06, UMR 7222, Institut des Systèmes Intelligents et de Robotique, F-75005, Paris, France

CNRS, UMR 7222, ISIR, F-75005, Paris, France
e-mail: raja.chatila@isir.upmc.fr

R. Hakli, J. Seibt (eds.), *Sociality and Normativity for Robots*, Studies in the Philosophy of Sociality 9, DOI 10.1007/978-3-319-53133-5_8

research on human-human joint action has become a topic of intense research in cognitive psychology and philosophy, providing elements and even offering control architecture hints to help our understanding of human-human joint action. We analyse some findings from these disciplines and connect them to the human-robot joint action case.

The work presented in this paper is a first necessary step toward the definition of an integrative framework needed for the design of autonomous robots that can engage in interaction with human partners. More precisely, we address the following questions:

- What knowledge does a robot need to have about the human it interacts with, and which processes does it need to handle to manage a successful interaction?
- Conversely, what information should the human possess to understand what the robot is doing and how the robot should make this information available to its human partner?

8.2 A Simple Scenario

We introduce a simple human-robot interaction scenario to illustrate the issues we address: a human and a robot have the common goal to build a stack with four blocks and to put a pyramid on the top of the stack. They are face to face. They should stack the blocks in a specific order. Each agent participates to the task by placing his/its blocks on the stack. At the end, one of the agents should place a pyramid on the top of the stack. The actions available to each agent are the following (with "object" = block or pyramid): take an object on the table, put an object on the stack, remove an object from the stack, place an object on the table, give an object to the other agent, support the stack (see next).

Figure 8.1 illustrates the initial state. Each agent can initially access only a subset of blocks and one of the two pyramids.

Each agent is able to perceive the state of the world and so knows where each object is, whether they can reach a given object, and can infer whether their partner can reach a given object. Moreover, we assume that each agent is able to observe the activity of the other. Figure 8.2 depicts the two possible final states.

A number of deviations from a nominal course are possible. For example, the stack might collapse or an agent might drop a block on their side of the table or on the opposite side. If the block falls on the opposite side, the question arises whether the other agent should put it directly on the stack or give it to the initial agent. Moreover, during the execution of the task, different types of behaviours are possible, including proactive behaviour (one agent could help the other one by supporting the stack while the other places a block on it), passive behaviour (one agent does not act at all) or incorrect behaviour (one agent does not stack blocks in the correct order or removes a correctly placed block from the stack).

Fig. 8.1 Initial state

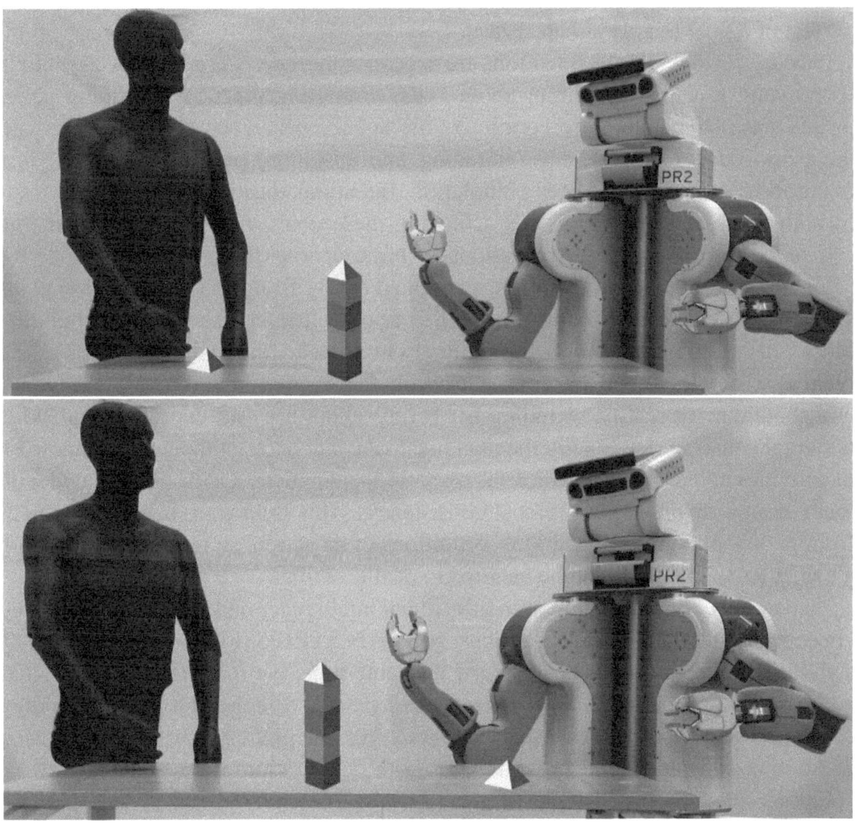

Fig. 8.2 Possible final states

The task needs to be set up at the beginning, for example through a dedicated command sent by the human to the robot. The way this is achieved is beyond the scope of this paper.

8.3 Acting Autonomously

Both philosophical and robotics approaches to joint action typically build from models of individual human or autonomous agents. Interestingly, existing approaches from both areas share a number of insights regarding the architecture needed to support individual action.

According to classical philosophical accounts of individual action, behaviour qualifies as action only if it has a certain type of mental antecedent or involves certain types of psychological processes (e.g. Davidson, 1980; Mele, 1992; Searle, 1983). Typically, this mental antecedent is identified as an intention. Intentions are often characterized as plans of action the organism chooses and commits itself to in pursuit of a goal (e.g. Bratman, 1987).

According to this view, intentions are executive attitudes whose functions include terminating practical reasoning about ends, prompting practical reasoning about means and plans, helping to coordinate the agent's behaviour over time and with the behaviour of other agents, initiating and sustaining intentional action, and monitoring and guiding it until completion. Intentions thus include representations of both goals and means towards achieving these goals, that is, action plans that can range from simple representations of basic actions to complex strategies for achieving distant goals. In addition, as pointed out by Bratman (1987), action plans are subject to rationality constraints. The various elements that form the building blocks of an action plan must be mutually consistent (internal consistency). The plan as a whole should be consistent with the agent's beliefs about the world and about current reality, including her beliefs about her own capacities and skills (external consistency). Finally the plan must take into account the wider framework of activities and projects in which the agent is also involved and be coordinated with them in a more global plan (global consistency). This brief overview of references from philosophy provides working definitions of an action, an intention, a goal and a plan, elements that should be handled to enable acting.

In parallel to this work, research in artificial intelligence and Robotics has defined concepts for autonomous agent actions, such as in STRIPS (Fikes & Nilsson, 1971), also based on means-ends analysis. At the same time, the robotics community has addressed the problem of *robot control architectures*, with the objective of building consistent and efficient robot system structures integrating perception, decision and action capacities, and providing for both deliberation and reactivity. Several solutions were proposed. One of them, which is commonly used today, is the three-layered architecture (Gat, 1992; Alami, Chatila, Fleury, Ghallab, & Ingrand, 1998; Muscettola, Nayak, Pell, & Williams, 1998; Nesnas, Wright, Bajracharya, Simmons, & Estlin, 2003; Saridis, 1995; Tambe, 1997), which defines:

- A *decision level*, which includes the capacities for producing a plan to accomplish a task and for supervising its execution, while being at the same time reactive to events from the next level below. The coexistence of these two features, a time-consuming planning process, and a time-bounded reactive supervisory process raises the key problem of their interaction and their integration to balance deliberation and reaction at the decisional level. Basically, the supervisory component uses the planner which may include temporal reasoning as a resource when needed and feeds the next level with the sequence of actions to be executed.
- An *execution control level*, or executive, which controls and coordinates the execution of functions distributed in operational modules (next level) according to the task requirements to achieve the plan. It is at this level that context-based action refinement is performed.
- A *functional level*, which includes all the basic built-in robot action and perception capacities. These functions are encapsulated into controllable communicating modules that enable the implementation of data processing and motor control loops (image processing, obstacle avoidance, motion control, etc.). In order to make this level as hardware independent as possible, and hence portable from a robot to another, it is connected with the sensors and effectors through a logical robot interface, i.e., an abstraction of these physical devices.

This architecture relies on representations of actions, goals, plans as well as robot's knowledge and skills. Building these representations remains an active research issue.

Interestingly, Pacherie (2008, 2012) proposes a dynamic model of intentions that also distinguishes three main stages in the process of action specification:

- A *distal intentions* level (D-intentions) in charge of the dynamics of decision making, temporal flexibility and high level rational guidance and monitoring of action;
- A *proximal intentions* level (P-intentions) that inherits a plan from the previous level and whose role is to anchor this plan in the situation of action, this anchoring has to be performed at two levels: temporal anchoring and situational anchoring;
- A *motor intentions* level (M-intentions), which encodes the fine-grained details of the action (corresponding to what neuroscientists call motor representations), is responsible for the precision and smoothness of action execution, and operates at a finer time scale than either D-intentions or P-intentions.

This suggests an interesting convergence between a philosophical account of the structure and dynamics of human action and a robot control architecture dedicated to action. From this, it appears relevant to consider whether a similar convergence could be established with regard to joint action.

8.4 Coordination Requirements in Joint Action

Successful joint action depends on the efficient coordination of participant agents' goals, intentions, plans, and actions. In other words, it is not enough that agents have a common goal and that then each set their own sub-goals, devise their own individual action plan and execute this plan. They must also *coordinate* their own sub-plans with those of their co-agents so as to have a coherent joint action plan and they must coordinate their actions during the *execution* phase to insure the successful completion of the joint action. For that they must monitor their partner's intentions and actions, predict their consequences and use these predictions to adjust their sub-plans, or in the execution phase, what they are doing to what their partners are doing. These processes, however, also play an important role in competitive contexts. In a fight, for instance, being able to anticipate the opponent's moves and to act accordingly is also crucial. A further requirement in the case of joint action is that co-agents share a goal and understand the combined impact of their respective intentions and actions on their joint goal and adjust them accordingly. In other words, agents should be able to align their representations of what they themselves and their partners are doing, and of how these actions together contribute to the shared goal.

As Michael and Pacherie (2015) point out, various forms of uncertainty can undermine mutual predictability, the alignment of representations and hence coordination. They include:

Motivational uncertainty: we might be unsure how convergent a potential partner's interests are with our own interests and thus unsure whether there are goals we share and can promote together. Additionally, even if we know what their current preferences are and that they match ours, we might be unsure how stable these preferences are.

Instrumental uncertainty: even assuming that we share a goal, we might be unsure what plan to follow to achieve that goal, or, if we have a plan, we might be unsure how roles should be distributed among us, or, even if the plan and the distribution of roles are settled, we might be uncertain when and where we should act.

Common ground uncertainty: we might be unsure how much of what is relevant to our deciding on a joint goal, planning for that goal and executing our plan is common ground, or mutually manifest to us. In other words, it is not sufficient, to ensure coordination, that we are actually motivated to pursue the same goals and have sufficiently similar instrumental beliefs and plans regarding how these goals should be achieved. We must also know or believe that this is the case.

These coordination constraints apply both to human-human joint action and to human-robot joint action and they can undermine both of them. However, they do not apply with the same strength. In human-human joint action, a human faces another human. The fact that they are both humans brings lots of shared background knowledge and assumptions can be made from both sides on what the other knows or not. This is far from easy to assess in the human-robot case. In this latter case, alignment processes need to be considered carefully to ensure an acceptable level of mutual predictability. On the robot side, this indicates that we need to integrate into the robot means to share representations explicitly with the human but also means to recognize and understand them (and to learn them if needed). On the other side, a human interacting with a robot is often disconcerted because it is difficult for him to have correct intuitions about robot capabilities or inabilities and perception abilities or weaknesses. To deal with this issue, some propose to train the human to use the robot, as we do for other technological devices (Cakmak & Takayama, 2014).

In what follows, we will first consider what resources humans can exploit in order to reduce uncertainty and achieve coordination at the level of intentions and action planning as well as at the level of action execution. To do that we will draw on recent conceptual and empirical work investigating the cognitive processes by which coordination in joint action is achieved. We will then consider human-robot joint action and the specific challenges it raises in addition to the challenges common with human-human joint action.

8.5 Coordination Processes in Human-Human Joint Action

Successful joint action requires agents to coordinate both their intentions and their actions. There has been a great deal of work in recent years, both conceptually and empirically, investigating the cognitive processes by which uncertainty is reduced and coordination achieved. Philosophical accounts of joint action have tended to concentrate on the conceptual requirements for shared intentions and to emphasize high-level action planning prior to acting. They are thus essentially concerned with the characterization of shared distal intentions. In contrast, cognitive psychology studies of joint action have explored the perceptual, cognitive, and motor processes that enable individuals to flexibly coordinate their actions with others online. The processes they describe are thus essentially processes involved in the formation and operation of shared proximal intentions and coordinated motor intentions. Because philosophers and psychologists focus on processes of uncertainty reduction that operate at different levels of action specification, it is important to bring together their complementary perspectives to shed light on the whole range of processes involved in acting together.

Philosophical accounts of shared intentions are attempts to cash out what it takes for agents to act in a jointly intentional manner. These accounts typically agree that shared (distal) intentions are more than mere summations of individual intentions. They agree therefore that something more is needed, although they tend to disagree on what more is needed. Rather than trying to adjudicate between different accounts, we take here the plurality of accounts as evidence that shared distal intentions may take different forms and be arrived at in a variety of ways.

According to Michael Bratman's very influential account (Bratman, 2014), shared intentions are characterized by a form of mutual responsiveness of each to each in their relevant intentions and plans. Responsiveness in intention means that each will adjust his subsidiary intentions concerning means and preliminary steps to the subsidiary intentions of others in a way that keeps track of the intended end of the joint action. It is thus essentially a matter of responsiveness in planning. Bratman describes negotiation, bargaining, shared reasoning and shared deliberation as some of the central processes through which mutual responsiveness in intentions is achieved.

Other philosophers have emphasized the essential role of joint commitments in joint actions. Thus, according to Margaret Gilbert (2009, 2014), joint commitments constitute the core of shared intentions: agents share an intention to do *A* if and only if they are jointly committed to intend as a body to do *A*. In the basic case, a joint commitment is created when each of two or more people openly expresses his personal readiness jointly with the other to commit themselves in a certain way, and it is common knowledge between them that all have expressed their readiness. According to Gilbert, these commitments have social normative force: participants in a joint activity have obligations towards each other to act in conformity with their shared intentions and correlative entitlements or rights to others so acting.

Finally, Raimo Tuomela (2007) points out that when agents act jointly as members of a group, what he calls we-mode joint action, they are often committed not just to a particular joint goal but also to a set of values, standards, beliefs, practices, social coordination conventions, pre-established scripts and routines and so on, that form the ethos of the group. The group ethos may thus serve to minimize uncertainty in joint actions.

While these philosophers have divergent views regarding the nature of the social glue that binds together the intentions of individuals in joint action (practical rationality for Bratman, the social normativity of joint commitments for Gilbert and collective acceptance of a group ethos for Tuomela), their accounts tend to be cognitively demanding: the coordination processes involved in forming and maintaining a shared intention rest on advanced representational, conceptual and communicational skills and sophisticated forms of reasoning about the complex interplay between each other's individual beliefs and intentions and the shared goal, about the mutual obligations and entitlements the shared intention generates, or about its relations to the group ethos.

In contrast to philosophical approaches, cognitive psychology studies of joint action have tended to focus not on the conceptual requirements for shared intentions but rather on the perceptual, cognitive, and motor processes that enable individuals to flexibly coordinate their actions with others online. Following Knoblich and colleagues (Knoblich, Butterfill, & Sebanz, 2011), we can distinguish between two broad categories of online coordination processes: emergent and intentional.

In intentional coordination, agents plan their own motor actions in relation to the joint goal and also to some extent to their partners' actions. As emphasized by Knoblich et al. (2011), shared task representations play an important role in goal-directed coordination. Shared task representations do not only specify in advance what the respective tasks of each of the co-agents are, they also provide control structures that allow agents to monitor and predict what their partners are doing, thus enabling interpersonal coordination in real time. Empirical evidence shows that having shared task representations influences perceptual information processing, action monitoring, control and prediction during the ensuing interaction (Heed, Habets, Sebanz, & Knoblich, 2010; Schuch & Tipper, 2007; Sebanz, Knoblich, Prinz, & Wascher, 2006). Thus, for instance, people tend to predict the sensory consequences not only of their own but also of other participants' actions (Wilson & Knoblich, 2005) and to automatically monitor their own and others' errors (van Schie, Mars, Coles, & Bekkering, 2004). Furthermore, several studies have shown that actors may form shared representations of tasks quasi-automatically, even when it is more effective to ignore one another (Atmaca, Sebanz, Prinz, & Knoblich, 2008; Sebanz, Knoblich, & Prinz, 2005; Tsai, Kuo, Hung, & Tzeng, 2008).

An important complement to the co-representation of tasks and actions is the co-representation of perception. In particular, joint attention provides a basic mechanism for sharing representations of objects and events and thus for creating a perceptual common ground in joint action (Tomasello & Carpenter, 2007; Tollefsen, 2005). Joint attention can also allow agents to perform joint actions more efficiently. For instance, a study by Brennan and colleagues (Brennan, Chen, Dickinson, Neider, & Zelinsky, 2007) demonstrated that co-agents in a joint visual search task were able to distribute a common space between them by directing their attention depending on where the other was looking and that their joint search performance was thus much more efficient than their performance in an individual version of the search task.

Another type of process that may contribute to better online coordination can be captured with the term "coordination smoother", i.e. any kind of modulation of one's movements that "reliably has the effect of simplifying coordination" (Vesper, Butterfill, Knoblich, & Sebanz, 2010, p. 2). For example, one may exaggerate one's movements or reduce variability of one's movements to make them easier for the other participant to interpret (Pezzulo, 2011). Although coordination smoothers may in some cases be produced automatically, the term may also be applied to processes, such as nods, winks and gestures, which are produced intentionally. And of course, there are a myriad other ways in which intentional alignment processes can reduce uncertainty, linguistic communication during the action being the paradigmatic case.

In emergent coordination, coordinated behaviour occurs due to perception-action couplings that make multiple individuals act in similar ways. One source of emergent coordination involves interpersonal entrainment mechanisms. For instance, people sitting in adjacent rocking chairs will tend to synchronize their rocking behaviour, even if the chairs have different natural rocking tempos (Richardson, Marsh, Isenhower, Goodman, & Schmidt, 2007). The perception of common or joint affordances can also lead to emergent coordination. A joint affordance is a case where an object affords action to two people that is may not afford to each of them individually. Thus, a seesaw may afford action to two kids, but not to a single child. A third source of emergent coordination is perception-action matching, whereby observed actions are matched onto the observer's own action repertoire and can induce the same action tendencies in different agents who observe one another's actions (Jeannerod, 1999; Prinz, 1997; Rizzolatti & Sinigaglia, 2010). It is likely that such processes make partners in a joint action more similar and thus more easily predictable, and thereby facilitate mutual responsiveness in action. Importantly, however, emergent forms of coordination can operate independently of any joint plans or common knowledge, which may be altogether absent, and do not ensure by themselves that the agents' actions track a joint goal.

Humans thus have at their disposal a vast array of coordination tools and processes, ranging from advanced representational, conceptual and communicational skills and sophisticated forms of reasoning to intentional and automatic online alignment processes, that they can use to reduce motivational, instrumental and common ground uncertainty and to promote interpersonal coordination. To enable efficient joint action, these processes must work together, as there are complementary limits on what each can do.

We must now examine whether, and under what conditions, these processes could play a similar role in human-robot interactions. It is important to note that some redundancy is present in the human case, as several combinations of these processes can be used to achieve the coordination required for successful joint action. Given that humans might have different expectations regarding a robot's capacities and given that the specificities of robotic cognitive architectures compared to human cognitive architectures may induce different cost/efficiency ratios in the use of these processes, the question also arises whether these processes should be deployed in different ways in human-robot interactions.

8.6 A (Tentative) Translation of Coordination Processes in Human-Robot Joint Action

We've seen that joint action presupposes the sharing of information at different levels, from object representations to task, action, intention and goal representations through the use of several processes. We will analyse now how such processes can make sense in a human-robot case and what kind of capacities they presuppose either on the robot or on the human side.

A pre-requisite of these processes is self-other distinction. As raised by Pacherie (2012, p. 359), it is important that agents be able to keep apart representations of their own and of others' actions and intentions.

On the robot side, this means that it should be able to handle a representation of itself and a representation of the human it interacts with, i.e., it must maintain a "mental" model of itself and a "mental" model of the human it interacts with. It should also be capable of updating these "mental" models as the action unfolds and the representations of the agents evolve. This in turn requires perspective-taking abilities, since the representations of the agents may evolve differently depending on their respective points of view.

On the human side, we can assume that the agent is able to handle several "mental" models. However, questions can be asked: does the human need to handle representations of robots' actions and intentions, in fact does he create such representations when interacting with a robot? Do we have to make it explicit at the beginning of a human-robot interaction that a robot makes use of actions and intentions representations (and which ones) to encourage its human partner to infer them?

Equipped with self-other distinction ability, the robot and its human partner need to understand what the other perceives (or does not perceive). More precisely, they must share knowledge about their interaction space. It is necessary that both the robot and the human identify objects to be acted upon, their location as well as location of possible obstacles. Thus they track the same objects and features of the situation and are mutually aware that they both do so. Here, *joint attention* is key because if joint attention is established, whatever information I can get, I can consider my partner would have it too if it occurs in the joint attention space. The interaction space includes what both partners perceive, but also what only one partner perceives (e.g., if one part of the table is hidden to the robot, the robot can establish that it cannot see a part of the environment but that the human can see it. Conversely, the robot can assume the human knows that a part of the table he can see is not visible to the robot).

This means that each agent must be equipped with situation assessment abilities that will enable them to anchor the situation of action (and this is in itself a complicated matter for the robot). Then, when acting jointly, each must ensure that they track the same objects and features of the situation as their partner. On the robot side, this means that the robot must have (necessarily partial) access to the human model of the real situation. On the human side, this means that the robot perception abilities should be readable by the human to enable him or her to draw inferences about what the robot perceives or not (noting that robot sensing abilities are not always easy to decode). Finally, both the robot and the human must be aware of that, so they both understand what are their perception capacities (and limitations).

This raises a number of questions: how can a robot know that the human it interacts with attended with him to the joint task? What are the cues that should be collected to infer joint attention? Symmetrically, how can a robot exhibit joint attention? What cues should the robot exhibit to let the human infer that joint attention is achieved? Moreover, once joint attention is achieved (or at least a given level of joint attention if we consider it is not a 0/1 question), how should it be managed during the overall course of joint action? How can we handle cooperative perception between a robot and a human and thus create perceptual common ground? Is there a need to negotiate about what should be jointly attended (or not)?

Another capacity, emphasized, among others, by Tomasello, Carpenter, Call, Behne, and Moll (2005) as a prerequisite to joint action, is *understanding intentional action*. Each agent should be able to read its partner's actions. To understand an intentional action, an agent should be able, when observing a partner's action or course of actions, to infer their partner's intention (i.e. their goal and plan). They should be able to exploit cues exchanged and to understand what their partner is attending to in their perceptive field.

That means that the robot needs to be able to understand what the human is currently doing and to be able to predict the outcomes of the human's actions.

To do so, it must be equipped with action recognition abilities (again potentially constrained to the current situation) and predictive action models enabling it to predict the outcomes of both its and the human partner's actions.

Complementarily, the human should be able to understand what the robot is currently doing and to predict the outcomes of robots' actions. To do so, he must be able to infer what is the underlying action when observing the robot's movement and to predict its outcome.

This process could be helped by the use of coordination smoothers. We can imagine that coordination smoothers could be added to already existing movement to facilitate this understanding. A human interacting with a robot would perhaps exaggerate her/his movement amplitude or do her/his movement exactly in front of the robot's dedicated perception sensor to ensure a good perception and understanding of his move. On the other side, the development of human-aware robot motion planning, that takes into account not only safety and efficiency but also legibility and social norms at planning level, could be considered as a software instance of coordination smoothers.

Equipped with *self-other distinction*, *joint attention* and *intentional action understanding* abilities, our agents should be able to understand actions in their perceptual context but this context should be enlarged to include the task and the joint goal to get the overall picture and allow coordination. This is where *shared task representations* come on stage. Equipped with such representations, our agents would be able not only to understand what the other is doing but also to predict what he/it will do next, e.g. by the use of action-to-goal or goal-to-action predictions. These predictions would help to make the entire interaction space more foreseeable. It enables also each agent to adapt his/its behaviour by taking into account this knowledge.

If we paraphrase the definition of Knoblich et al. (2011), this means that we must equip the robot with a model of the respective tasks of each of the co-agents and also with control structures that will allow it to monitor and predict what its partners are doing. On the other side, the human must be aware of the respective tasks of each of the co-agents and how to monitor them. Doing that can be considered as putting in perspective all the processes already described. For example, joint attention allows to know that both agents track the same object, intentional action understanding allows to infer that the robot is currently moving this object to a goal position, and the existence of a shared task representation enables to interpret the action as a contribution to the common goal.

It is important to point out what it means to share information in terms of information alignment. The robot and the human need to understand, to interpret the world in the same way, their understandings/interpretations need to be aligned at some point and this is a component of the ability to share. It is crucial for enabling coordination and communication among the agents.

For example, it is not sufficient for the robot to perceive that a blue object stands at position (x, y, z); it must know what object it is in order to be able to share this information with the human. It is not sufficient for the robot to interpret an arm movement as "something moves in front of me"; rather, it must interpret it as "the human hands an object to me" so it can react accordingly to this action.

This concern needs also to be taken into account on the human side. The human should be aware of the limitations of the representational resources of the robots to avoid over-interpretations by humans of what the robot knows and understands about the scene.

This alignment issue can also be considered from a broader viewpoint. Tuomela (2007) states that the involved agents should share what he calls group ethos, Tomasello et al. (2005) speak about cultural creation/learning, and Clark (1996) about common ground. We have already considered the set of information that needs to be shared to handle a joint action, but here the spectrum is larger. It concerns the set of values, standards, beliefs, practices, social coordination conventions, pre-established scripts and routines. How is it possible to model such concepts in a robot and to what extent could we consider that they are shared by the human and the robot?

Another philosophical account concerns joint commitment. According to Gilbert (2014) a shared intention to perform a joint action essentially involves a joint commitment on the part of the co-agents, where a joint commitment creates a set of mutual obligations and entitlements for agents to perform their part in the joint endeavour. How can a robot and a human express and share this kind of engagement? How can they express their readiness to be jointly committed in the first place? How do they monitor whether or not they stay committed (or not) as the action unfolds? How is the joint commitment terminated once the common goal is achieved? How are such commitments represented and updated?

Finally, Bratman (2014) proposes that shared intentions are characterized by mutual responsiveness. Engaged in a joint action, agents need to be able to share not just representations, but also reasoning processes toward the joint goal, to be able to deliberate together, to negotiate. Such processes required high level reasoning abilities. How can a robot and a human reason together toward their joint goal? What are the reasoning abilities that need to operate? Which media could be used to enable such reasoning? How to model it? We propose in Fig. 8.3, a representation of knowledge and processes a robot need to handle to operate a joint action with a human.

8.7 A Framework for Joint Action

As stated by Knoblich et al. (2011), philosophers generally agree that "what distinguishes joint actions from individual actions is that the joint ones involve a shared intention and shared intentions are essential for understanding coordination in joint action" (p. 60). Tomasello et al. (2005) say nothing else when they claim that "Understanding the intentional actions and perception of others is not by itself sufficient to produce humanlike social or cultural activities. Something additional is required. Our hypothesis for this 'something additional' is shared intentionality".

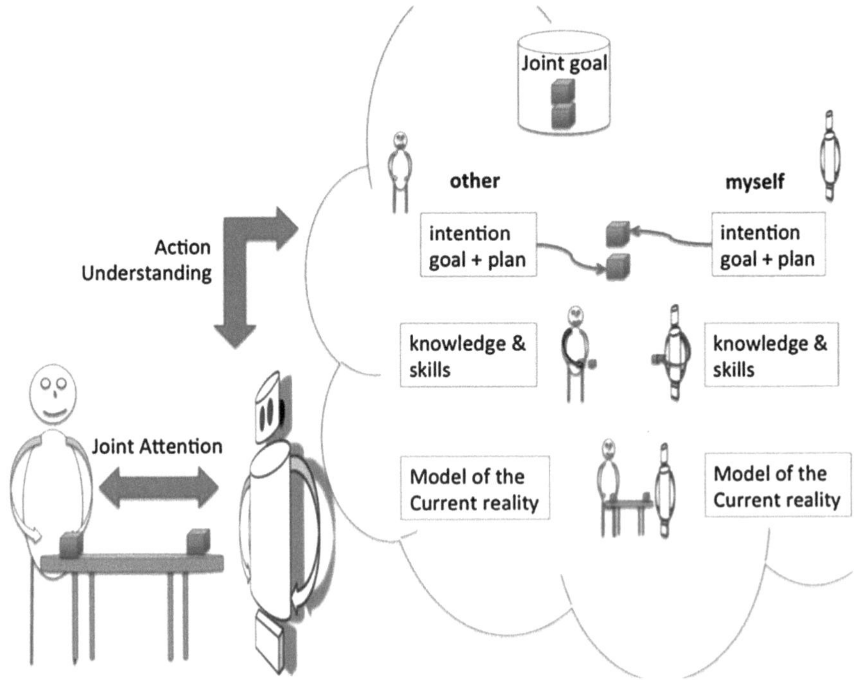

Fig. 8.3 Knowledge and processes for joint action. The robot builds and maintains a distinct mental model of itself and of its human partner concerning the state of the world. It also reasons and builds its own behaviour based on its estimation of its human partner intentions, knowledge and skills

And they add: "shared intentionality refers to collaborative interactions in which participants have a shared goal (shared commitment) and coordinated action roles for pursuing that shared goal".

Successful joint action depends on the efficient coordination of participant agents' goals, intentions, plans, and actions. In other words, in joint action, it is not enough that agents control their own actions, i.e., correctly predict their effects, monitor their execution and make adjustments if needed. In addition, they must also coordinate their intentions and actions with those of their co-agents so as to achieve their joint goal.

It has to be noticed that AI community has proposed seminal work on teamwork such as Cohen and Levesque (1991) and Grosz and Kraus (1996). What we propose here is to analyse Pacherie's (2007, 2011, 2012) theory of joint action, which also considers three levels of action. If we try to map this theory to robot architecture, we can describe these three levels as the following: a shared distal/decisional level, a shared proximal/execution level and a coupled motor/functional level.

8.7.1 Shared Distal/Decisional Level

At the shared distal/decisional level, acting alone, the robot handles its goal, plan and decision-making; all elements that it represents would be realized by itself. Acting jointly, the robot must be able to handle joint goal, plan and action representation and possibly cooperative decision-making (including, e.g., joint planning abilities). It will represent not only what would be achieved by itself but also by the other (with potentially different levels of granularity and completeness). Moreover, high level monitoring would include not only the robot's monitoring of its own actions and goals but also more generally monitoring of the joint goal and consequently monitoring of the other actions too.

Pacherie (2012) explains that at this level, the participating agents (1) represent the overall goal yet need not represent the whole plan but only their own sub-plans and the meshing parts of the sub-plans of others and (2) some of what they represent is to be performed by others. Agents need to be able to handle triadic and dyadic adjustment at that level.

8.7.2 Shared Proximal/Execution Level

It is at the shared proximal/execution level that situational and temporal anchoring of the action take place, which means that the action plan inherited from the distal/decision level must be further refined and adjusted to the situation at hand in order for the action to be launched and its unfolding monitored and controlled. At that level, the robot and the human need to be able to share representations (in the best case jointly) and to coordinate their perceptions (to achieve joint attention) in order to coordinate their actions and possibly realize adjustment (dyadic, triadic and collaborative) in the current context.

Pacherie (2012) explains that for agents to share a proximal intention, the following should hold: (1) agents each represent their own actions and their predicted consequences in the situation at hand *(self-predictions)*, (2) agents each represent the actions, goals, motor and proximal intentions of their co-agents and their consequences *(other-predictions)*, (3) agents each represent how what they are doing affects what others are doing and vice-versa and adjust their actions accordingly *(dyadic adjustment)*, (4) agents each have a representation (which may be only partial) of the hierarchy of situated goals and desired states culminating in the overall joint goal *(joint action plan)*, (5) agents each predict the joint effects of their own and others' actions *(joint predictions)*, and (6) agents each use joint predictions to monitor progress toward the joint goal and decide on their next moves, including moves that may involve helping others achieve their contributions to the joint goal *(triadic adjustment)*.

That means the robot needs to be able to handle: its world representation, a world representation of the human it interacts with (potentially limited to the task to be performed), the possible effect of its actions on the human actions (and vice versa), their joint goal and action plan representation, a prediction of their actions, a means to monitor progress toward the joint goal (and possibly a means to revise the on-going joint plan). A triadic adjustment means that the robot and the human can adapt their behaviour toward the joint goal. This implies, for example, that if the human drops his object in the robot space, the robot will place the object on the stack. If it had done a dyadic adjustment it would have made the object accessible to the human to let him finish the action. A dyadic adjustment means that the robot and the human can adapt their behaviour to the other's actions (not toward the joint goal) Interestingly, Tomasello et al. (2005) have proposed that the capacity for triadic engagement presents two phases in the course of human development. At around 9–12 months of age, infants begin to interact together with a goal-directed agent toward some shared goal. In doing this, both perceptually monitor the behaviour and perceptions of their partners with respect to that shared goal. However, it is only at around 12–15 months of age, that they begin to engage in significant amounts of coordinated joint engagement, understanding not just the shared goal but also beginning to understand the complementarity between their own and their partner's specific action plans. As Tomasello and colleagues point out: "This means, for instance, that the child understands that in pursuing the shared goal of building a block tower the adult holds the edifice steady while she, the child, places blocks. Infants of this age not only share goals but also coordinate roles" (Tomasello et al., 2005, p. 682). This understanding thus makes possible more flexible triadic adjustment processes, such as reversing roles with a partner or helping the partner play his role. These adjustment mechanisms exploit shared task representations that not only specify in advance what the respective tasks of each of the co-agents are but also provide control structures that allow for flexible coordination.

8.7.3 Coupled Motor/Functional Level

The coupled motor/functional level corresponds to robot sensory-motor behaviour that would allow to achieve high-bandwidth interaction with the human partner. An example could be exchanging an object with a human and the associated force-feedback processes. In such tight situations involving precise coordination between the actors, the parameterization of the functional level needs to be coupled with the one of the other actor. This means that the robot control loops would be directly parameterized by the other actor's motions or actions.

8.8 Conclusion: Toward a Framework for Joint Action

In this paper we proposed an analysis of some findings in psychology and philosophy in the domain of human-human joint action. Our aim was to identify knowledge, representations and processes that a robot, interacting with a human, needs to possess and exploit. Complementarily, we analysed what information needs to be shared with the human to enable a consistent interaction. We have seen that, as already pointed out by cognitive psychology and philosophy, self-other distinction, joint attention, intentional action understanding and shared task representations as well as common ground, joint commitment and mutual responsiveness make sense in our context. We came up with a set of questions about their management in our context.

We then tried to apply the framework proposed by Pacherie (2012) to a human-robot case and show that it could fit the development of an architecture dedicated to human-robot interaction. It is inspiring in the search to frame an architecture dedicated to human-robot interaction. We show that this three-layer division seems meaningful not only for the robot, human and the human-human cases but also for the human-robot case.

This paper is a first step toward the objective of identifying and describing precisely the different robot abilities and how they are involved in the overall process of collaborative human-robot task achievement. To this end, we placed ourselves purposefully at a conceptual level. This analysis is obviously sustained by the work of the human-robot interaction community to which we contribute. We refer the interested reader to Lemaignan, Warnier, Sisbot, Clodic, and Alami (2016) and Kruse, Pandey, Alami, and Kirsch (2013) where we discuss a number of such abilities studied and implemented by robotics researchers.

Acknowledgements This work has been funded by the French Agence Nationale de la Recherche ROBOERGOSUM project ANR-12-CORD-0030.

References

Alami, R., Chatila, R., Fleury, S., Ghallab, M., & Ingrand, F. (1998). An architecture for autonomy. *The International Journal of Robotics Research, 17*(4), 315–337.

Atmaca, S., Sebanz, N., Prinz, W., & Knoblich, G. (2008). Action co-representation: The joint SNARC effect. *Social Neuroscience, 3*(3–4), 410–420.

Bratman, M. E. (1987). *Intention, plans, and practical reason*. Cambridge: Harvard University Press.

Bratman, M. E. (2014). *Shared agency: A planning theory of acting together*. Oxford: Oxford University Press.

Brennan, S. E., Chen, X., Dickinson, C., Neider, M., & Zelinsky, G. (2007). Coordinating cognition: The costs and benefits of shared gaze during collaborative search. *Cognition, 106*, 1465–1477.

Cakmak, M., & Takayama, L. (2014). Teaching people how to teach robots: The effect of instructional materials and dialog design. In *Proceedings of the 2014 ACM/IEEE International Conference on Human-Robot Interaction*, Bielefeld (pp. 431–438). ACM.
Clark, H. H. (1996). *Using language*. Cambridge: Cambridge University Press.
Cohen, P. R., & Levesque, H. J. (1991). Teamwork. *Nous, 25*(4), 487–512.
Davidson, D. (1980). *Essays on actions and events*. Oxford: Oxford University Press.
Fikes, R. E., & Nilsson, N. J. (1971). STRIPS: A new approach to the application of theorem proving to problem solving. *Artificial intelligence, 2*(3–4), 189–208.
Gat, E. (1992). Integrating planning and reacting in a heterogeneous asynchronous architecture for controlling real-world mobile robots. In *Proceedings of the Tenth National Conference on Artificial Intelligence*, Menlo Park (pp. 809–815). AAAI Press.
Gilbert, M. (2009). Shared intention and personal intentions. *Philosophical Studies, 144*, 167–187.
Gilbert, M. (2014). *Joint commitment: How we make the social world*. Oxford: Oxford University Press.
Grosz, B. J., & Kraus, S. (1996). Collaborative plans for complex group action. *Artificial Intelligence, 86*(2), 269–357.
Heed, T., Habets, B., Sebanz, N., & Knoblich, G. (2010). Others' actions reduce crossmodal integration in peripersonal space. *Current Biology, 20*, 1345–1349.
Jeannerod, M. (1999). The 25th bartlett lecture. to act or not to act: Perspectives on the representation of actions. *Quarterly Journal of Experimental Psychology, 52A*, 1–29.
Knoblich, G., Butterfill, S., & Sebanz, N. (2011). Psychological research on joint action: Theory and data. *Psychology of Learning and Motivation-Advances in Research and Theory, 54*, 59–101.
Kruse, T., Pandey, A. K., Alami, R., & Kirsch, A. (2013). Human-aware robot navigation: A survey. *Robotics and Autonomous Systems, 61*(12), 1726–1743.
Lemaignan, S., Warnier, M., Sisbot, E. A., Clodic, A., & Alami, R. (2016, in press). Artificial cognition for social human-robot interaction: An implementation. *Artificial Intelligence*. http://www.sciencedirect.com/science/article/pii/S0004370216300790
Mele, A. R. (1992). *Springs of action*. Oxford: Oxford University Press.
Michael, J., & Pacherie, E. (2015). On commitments and other uncertainty reduction tools in joint action. *Journal of Social Ontology, 1*(1), 89–120.
Muscettola, N., Nayak, P. P., Pell, B., & Williams, B. C. (1998). Remote agent: To boldly go where no ai system has gone before. *Artificial Intelligence, 103*(1), 5–47.
Nesnas, I. A., Wright, A., Bajracharya, M., Simmons, R., & Estlin, T. (2003). CLARAty and challenges of developing interoperable robotic software. In *Proceedings of 2003 IEEE/RSJ International Conference on Intelligent Robots and Systems, 2003. (IROS 2003)*, Las Vegas (Vol. 3, pp. 2428–2435). IEEE.
Pacherie, E. (2007). Is collective intentionality really primitive? In M. Beaney, C. Penco, & M. Vignolo (Eds.), *Mental processes: Representing and inferring* (pp. 153–175). Cambridge: Cambridge Scholars Press.
Pacherie, E. (2008). The phenomenology of action: A conceptual framework. *Cognition, 107*(1), 179–217.
Pacherie, E. (2011). Framing joint action. *Review of Philosophy and Psychology, 2*(2), 173–192.
Pacherie, E. (2012). The phenomenology of joint action: Self-agency vs. joint-agency. In A. Seemann (Ed.), *Joint attention: New developments* (pp. 343–389). Cambridge: MIT Press.
Pezzulo, G. (2011). Shared representations as coordination tools for interaction. *Review of Philosophy and Psychology, 2*(2), 303–333.
Prinz, W. (1997). Perception and action planning. *European Journal of Cognitive Psychology, 9*, 129–154.
Richardson, M. J., Marsh, K. L., Isenhower, R. W., Goodman, J. R. L., & Schmidt, R. C. (2007). Rocking together: Dynamics of unintentional and intentional interpersonal coordination. *Human Movement Science, 26*, 867–891.
Rizzolatti, G., & Sinigaglia, C. (2010). The functional role of the parieto-frontal mirror circuit: Interpretations and misinterpretations. *Nature Reviews Neuroscience, 11*, 264–274.

Saridis, G. N. (1995). Architectures for intelligent controls. In M. M. Gupta & N. K. Sinha (Eds.), *Intelligent control systems: Theory and applications* (pp. 127–148). Piscataway, NJ: IEEE Press.
Schuch, S., & Tipper, S. P. (2007). On observing another person's actions: Influences of observed inhibition and errors. *Perception & Psychophysics, 69*, 828–837.
Searle, J. R. (1983). *Intentionality*. Cambridge: Cambridge University Press.
Sebanz, N., Knoblich, G., & Prinz, W. (2005). How two share a task: Corepresenting stimulus–response mappings. *Journal of Experimental Psychology: Human Perception and Performance, 31*, 1234–1246.
Sebanz, N., Knoblich, G., Prinz, W., & Wascher, E. (2006). Twin peaks: An ERP study of action planning and control in co-acting individuals. *Journal of Cognitive Neuroscience, 18*, 859–870.
Tambe, M. (1997). Towards flexible teamwork. *Journal of Artificial Intelligence Research, 7*, 83–124.
Tollefsen, D. P. (2005). Let's pretend: Children and joint action. *Philosophy of the Social Sciences, 35*(75), 74–97.
Tomasello, M., & Carpenter, M. (2007). Shared intentionality. *Developmental Science, 10*(1), 121–125.
Tomasello, M., Carpenter, M., Call, J., Behne, T., & Moll, H. (2005). Understanding and sharing intentions: The origins of cultural cognition. *Behavioral and Brain Sciences, 28*, 05.
Tsai, C. C., Kuo, W. J., Hung, D. L., & Tzeng, O. J. (2008). Action co-representation is tuned to other humans. *Journal of Cognitive Neuroscience, 20*(11), 2015–2024.
Tuomela, R. (2007). *The philosophy of sociality*. Oxford: Oxford University Press.
van Schie, H. T., Mars, R. B., Coles, M. G., & Bekkering, H. (2004). Modulation of activity in medial frontal and motor cortices during error observation. *Nature Neuroscience, 7*(5), 549–554.
Vesper, C., Butterfill, S., Knoblich, G., & Sebanz, N. (2010). A minimal architecture for joint action. *Neural Networks, 23*(8–9), 998–1003.
Wilson, M., & Knoblich, G. (2005). The case for motor involvement in perceiving conspecifics. *Psychological Bulletin, 131*(3), 460.

Chapter 9
Joint Commitments and Group Identification in Human-Robot Interaction

Alessandro Salice and John Michael

Abstract This paper investigates the possibility of designing robots that are able to participate in commitments with human agents. In the first part of the article, we tackle some features that, we claim, make commitments crucial for human-human interactions. In particular, we focus on some reasons for believing that commitments can facilitate the planning and coordination of actions involving multiple agents: not only can commitments stabilize and perhaps even increase the motivation to contribute to other agents' goals and to shared goals, they also reinforce agents' willingness to rely on other agents' contributions. In the second part, we turn our attention to human-robot interaction. Here, we elaborate on five problems that roboticists could encounter in the attempt to implement commitments in human-robot interactions, and we argue in favor of some possible solutions to those problems. Finally, in the last part of the paper we zoom in on joint commitments, i.e., on commitments held by a plurality of agents towards shared goals. Given that the concept of joint commitment invokes the notion of a group, we discuss some more specific challenges that would have to be met for human agents to group-identify with robots.

Keywords Commitment • Trust • Social robotics • Groups • Social identity • Human-robot interaction • Joint action • Cooperation • Coordination

9.1 Introduction: Robots and Commitments

Is it possible to design robots that could make commitments in which human agents would be willing to put their trust? And, conversely, is it possible to design

---

A. Salice (✉)
Department of Philosophy, University College Cork, 4 Elderwood, College Road, Cork, Ireland
e-mail: alessandro.salice@ucc.ie

J. Michael
Department of Philosophy, University of Warwick, CV4 7AL, Coventry, UK

Department of Cognitive Science, Central European University, Oktober 6 Utca 7, 1051 Budapest, Hungary
e-mail: michaelj@ceu.edu

R. Hakli, J. Seibt (eds.), *Sociality and Normativity for Robots*, Studies in the Philosophy of Sociality 9, DOI 10.1007/978-3-319-53133-5_9

robots so that human agents would be motivated to honor commitments made to them? This is not a purely speculative question. For the existence of robots has an impact on human society, and the–increasing–presence of robots may already be modifying the way in which we view social interactions, social relations and the social world. Against this background, it could be argued that it would be very useful to implement commitments in human-robot interactions. Such an implementation could, for example, enhance human agents' willingness to rely on robots (e.g. to perform actions which depend upon a contribution from a robot and otherwise entail a risk), or it could motivate human agents to perform important actions which they otherwise might not perform (e.g. to take their pills regularly because they have made a commitment to their robot nurse that they will do so, or more generally, to promote therapeutic interactions – especially with regard to elderly people, cf. Mordoch, Osterreicher, Guse, Roger, & Thompson, 2013). In our view, the consequences of implementing commitment, which is so crucial for human joint actions, in human-robot interaction could be of considerable importance, and could range over many different domains of human agency: from disaster relief to health care, from education to manufacturing, etc.

In this paper, we want to focus on several general challenges that are likely to arise in implementing commitments in human-robot interaction, and we aim at canvassing options for meeting these challenges (Sect. 9.2). In doing so, we will home in on one particular type of commitment, namely *joint* commitment. Because of the link between joint commitments and *group identification*, as we shall see, this kind of commitments poses several unique challenges, but also promises several unique benefits for human-robot interaction (cf. Sect. 9.3).

However, before moving to the specific case of commitments in human-robot interaction, it might be helpful to, first, highlight in what sense it can be argued that (explicit and implicit) commitments serve important functions which are of great benefit in *human-human* interactions. In the remainder of this section, we will lay out some conceptual preliminaries pertaining to the concept of commitment. In particular, we will set out a rough framework within which several different types of commitment can be distinguished, and specify the functions which commitments can serve.

The notion of commitment most philosophers are concerned with is *volitional commitment*. As Gilbert (2006) puts it, a commitment is a "creature of the will" in two important respects. First, a commitment is generated *by the will*. Second, a commitment is *of the will* in the sense that it binds the will. A commitment also has a content, and the fact that it is a creature of the will imposes constraints on what this content can be: one can only commit to things that fall within the purview of one's voluntary control. This means, on the one hand, that one can form a commitment to, say, lose 10 kg before summer, but one cannot commit to growing 10 cm taller. On the other hand, it also means that one can only commit to someone else's doing something to the extent that one holds control or authority over that person's behavior. Thus, a queen can commit to sending her own troops to a battle, but she cannot commit to sending some other queen's troops.

In addition to having authors, commitments also have recipients, i.e. the person or persons you commit to. One way to start constructing a typology of commitments is in terms of whom their *authors* or *recipients* are. Herbert Clark (2006) proposes such a typology, and here we use a slightly modified and expanded version of his typology. First, the author and the recipient of the commitment can be one and the same person (*self-commitment*), or they can be different people (*interpersonal commitments*). A self-commitment can be, e.g., the result of a private decision: if you decide to lose 10 kg before summer, you also commit to that. Examples of interpersonal commitments are commitments made to others, for example through promises: if I promise to have lunch with you tomorrow, I am committed to that towards you.

This first distinction is orthogonal to a second distinction: self-commitments and interpersonal commitments can be either *private* or *public*. A private commitment is a commitment known only to its author and recipient. In the case of private self-commitments, since the author is the same person as the recipient, this means that only one person knows of the commitment. A public commitment, in contrast, is a commitment that has an *audience*. For instance, Peter may commit to quitting smoking but tell no one (private self-commitment), or he may tell his family (the audience) about his commitment (public self-commitment). The key difference between audience and recipient is that a commitment creates obligations (and corresponding entitlements) vis-à-vis the recipient but not vis-à-vis the audience. Telling his family about his commitment to quitting smoking (public self-commitment) may create an additional motivation for Peter, who would incur the risk of teasing and public embarrassment, but his family, as a mere audience to the commitment, would not be entitled to his delivering on it. In contrast, if Peter were to commit to his family to exercise more (e.g. make a promise to them), he would have an obligation to them, and they could hold him responsible for his failing to fulfill his commitment.

In addition, public (but not private) commitments can also be *interdependent*. Among interdependent commitments, one can distinguish *unilateral* commitments (in which case one agent makes a commitment to a second agent, but the second agent is not committed to anything) from *mutual* commitments (in which case she is also committed to something). Furthermore, mutual commitments can be either *complementary* (as when Peter is committed to digging a hole as long as Jim is committed to paying him for it) or *joint* (Peter and Jim are committed to a shared goal, such as painting the house together). In the context of joint action, which will be our focus, it is this latter kind of commitment (i.e., joint commitment) that is most directly of interest.

Joint commitments are a pervasive element of human sociality. When human agents are performing a joint action together, it is highly typical for them to be jointly committed to that joint action. As a result, they have a sense that they ought to remain engaged until the shared goal is achieved or until all participants have agreed to interrupt the joint action. Thus, to borrow Margaret Gilbert's famous example, if two people have agreed to take a walk together, they are likely to take themselves to have obligations to walk at a pace that the other can keep up with, to refrain from suddenly veering off into the woods without giving any explanation, and to wait for the other if she has to tie her shoe (cf. Gilbert, 1989). Correspondingly, each of them will typically expect that the other will also fulfill these obligations, and otherwise take herself to be entitled to reprimand her. In the canonical case, this is because each has given the other an assurance that they will remain committed. However, even in the absence of *explicit* assurances, *implicit* commitments arising through bodily cues, subtle nuances of verbal communication or situational affordances may have very similar effects, and be experienced in very similar ways.

A consequence of commitments in general, if they are credible, is that they make agents' behavior *predictable* in the face of fluctuations in their desires and interests. Thus, they make it possible to have expectations that one would not otherwise have about events that may occur and that would be relevant to one's goals. This is true of self-commitments insofar as a commitment made to oneself to pursue a goal facilitates planning by ruling out action options which might be attractive but which are incompatible with the goal to which one is committed (Bratman, 1987). Moreover, by staking one's self-image (or reputation) to a self-commitment, one can motivate oneself to pursue that goal. In fact, one might speculate that this may be part of the reason why so-called "sunk costs" effectively motivate one to remain engaged in pursuance of goals in which one has already invested (Kahneman & Tversky, 1979).

But it is all the more true of interpersonal commitments, and in particular of joint commitments. Interpersonal commitments clearly have a very useful social function, in part because they make people willing to perform actions that they otherwise would not perform – to work, for example, given that somebody has made a commitment to pay them for it, or to lift one end of a heavy table that cannot be moved by one person alone. More specifically, having reliable expectations about others' actions facilitates cooperation and coordination. In cooperation problems, such as the prisoners' dilemma game (Axelrod, 1984), individuals are tempted to defect in order to maximize their own benefits but by cooperating maximize the overall group benefit. In coordination problems, on the other hand, such as the stag hunt game (Skyrms, 2004), two agents each maximize their individual benefits if they coordinate their actions, but get no benefit if they try to coordinate but fail to do so, and are therefore tempted to opt for a smaller benefit which does not depend upon coordinating with the other agent. Moreover, having reliable expectations about others' contributions to one's goals, or to shared goals, is also useful insofar as it may facilitate the planning of joint actions with complementary subplans, which depend upon and build upon each other, as well as the online coordination of joint actions among multiple agents (Pacherie, 2013; Vesper, Butterfill, Knoblich, & Sebanz, 2010).

In view of these clear social functions that come with *interpersonal* commitment, we believe that it is well worth considering whether and how interpersonal commitments may be implemented in human-robot interaction. (It should not go unmentioned at this stage that *self*-commitments have been implemented in robots, cf. Cohen & Levesque, 1990.) It seems already possible to design robots that could make commitments in which humans would be willing to put their trust, but under what conditions would humans be motivated to honor commitments made to robots (cf. Kahn et al., 2015)–i.e. to perform actions that they perceived no longer to be in their interests *because they had made a commitment to a robot?* Identifying these conditions could provide new possibilities for motivating people to perform important actions such as taking their medication, practicing proper hygiene, practicing new skills that they are learning, and putting in the requisite level of effort at manufacturing jobs.

In the following section (Sect. 9.2), we will therefore consider challenges to confront in implementing interpersonal commitment in general in human-robot interaction. Next (Sect. 9.3), we will turn to the more specific challenge of whether and how joint commitments can be implemented in human-robot interaction.

9.2 Interpersonal Commitment in Human-Robot Interaction Made Easy

One conceptual challenge arises immediately when considering how to design robots that can participate in commitments with humans. One might simply be unprepared to believe that robots are appropriate candidates for participating in commitments. Call this the "But robots can't commit!" objection (cf. Searle, 2010, 135). The intuition behind such an objection may be that robots are programmed to act in specific ways and therefore not free to value action options simply because they have made a commitment to do so. Commitments involve a resistance to the pull of outside options that would maximize one's interest. But how could a robot ever resist the action option that it is programmed to value most highly? On the other hand, if the robot is programmed simply to do what it has committed to doing, then one might doubt whether it is doing so because of the commitment rather than simply because it is programmed to favor action options to which it has committed.

Clearly, this objection taps very fundamental and complex issues (Is free will a prerequisite for making commitments? Do *humans* even have free will? Do commitments require the ability to understand norms? Etc.). Luckily, we believe it is both possible and productive to get around this objection by taking a pragmatic approach.

Our starting point is the observation that even if a human agent does not explicitly believe that a robot can make commitments, or that she herself owes it to a robot to honor a commitment that she has made to the robot, she may nevertheless implicitly sense the opposite. Indeed, there is evidence that people frequently confide in robots and experience their relationship with robots as being characterized by trust even if they do not explicitly believe that the robot is conscious or has emotions, etc. (Hollander & Wijnberg, 2009). More generally, we often act as though committed and trust other agents to do so even if we would not explicitly judge that a commitment really is in place. An example will help to illustrate this: Sam is cleaning up the living room and picks up a ball that had been lying on the floor. As it happens, his dog Woofer notices this and bounds over to him, apparently ready to play fetch. Sam was not intending to play fetch and does not particularly desire to, but may now feel obligated to, because he has generated an expectation on the part of Woofer that they will now play fetch together. In cases like this one, one may sense that a commitment is in place and act accordingly even if one would explicitly deny that there is any commitment. By the same token, one sometimes treats artifacts such as laptops and mobile phones as though they were committed to contributing to one's goals. For example, if Frank is counting on his mobile phone to provide him with the address of the party to which he is invited, but the mobile phone fails to do so (e.g. because the battery runs out or the file has been deleted), he may become angry at the mobile phone and feel as though it had betrayed him. Thus, the challenge to roboticists need not be (although it could be) to design robots that *really do* engage in commitments in the strict sense, or which people *explicitly believe* can engage in commitments in the strict sense, but to design robots that human agents tend to interact with as though such commitments were in play regardless of whether the human interactants explicitly judge that they are in play.

One might, on the one hand, formulate this point by saying that implementing "quasi-commitments" would be good enough in order to gain the potential social benefits associated with commitments. On the other hand, a different option is to conceptualize such cases (i.e. cases in which human agents feel and/or act committed, or feel and/or act as though a commitment were owed to them, even though they would not explicitly judge that a commitment is in place) to be continuous with cases of commitment in the strict sense. In order to flesh out this latter option, we draw upon a strategy developed by Michael, Sebanz and Knoblich (2016), which is based upon a characterization of the minimal structure of situations in which a subjective *sense of commitment* can arise. This minimal structure can be expressed as follows:

(i) There is an outcome which an agent (ME) either desires to come about, or which is the goal of an action which ME is currently performing or intends to perform. We will refer to this outcome as "*G*' (for "goal").
(ii) The external contribution (*X*) of a second agent (YOU) is crucial to bringing about *G*.

Clearly, conditions (i) and (ii) specify a broader category than that of commitment in the strict sense. Nevertheless, situations with this structure may elicit a sense of commitment on the part of one or both agents. Following Michael et al. (2016), we conceptualize the sense of commitment as follows:

> ME has a sense that YOU is committed to performing *X* to the extent that ME expects *X* to occur because (i) and (ii) obtain.
> YOU has a sense of being committed to performing *X* to the extent that YOU is motivated by her belief that ME expects her to contribute *X*.

While the minimal structure is specified such that only one agent (ME) desires *G* and/or has *G* as a goal, there are many cases in which both agents desire *G* and/or have the goal *G*. In those cases, the commitment may be mutual, with each agent having a sense of being committed as well as a sense that the other agent is committed.

Against this background, the challenge in implementing commitments in human-robot interaction, then, is transformed into a set of specific, tractable challenges – namely, to implement factors modulating the sense of commitment. In this vein, our aim in the rest of this section is to identify and evaluate several factors that may be useful for the project of implementing commitment in human-robot interactions. Again, it must be emphasized that the aim in implementing such factors need not be to design robots that people explicitly believe can engage in full-blown commitments in the strict sense. Rather, the primary aim of the pragmatic approach espoused here is to design robots with whom human agents tend to interact with a sense of commitment, irrespectively of whether they explicitly judge that commitments are in play, and regardless of whether commitments in the strict sense are in play.

The set of factors that we will be discussing is by no means exhaustive – as we have already noted, the minimal approach allows for a broad range of factors to qualify as modulating the sense of commitment. Instead, we have sought to identify five factors that may be especially useful for implementing joint commitments in human-robot interactions.

Some of the factors and mechanisms of how we communicate and coordinate social interactions with robots are well-known to roboticists and already identified as design principles (cf. Breazeal, 2004). The novel element added here is that these factors are systematized as contributions to the creation of a sense of commitment in HRI.

9.2.1 Reputation

One factor which is likely to play an important role in determining whether an agent is likely to honor her commitments is one's prior knowledge of that agent, either from 2nd person interactions with her or from third-party information about that agent (i.e. reputation).

Intriguingly, humans' sensitivity to agent identity as a cue to trustworthiness is not limited to human-human interactions: 6–10 month old human infants who observe animated figures interacting with each other tend to prefer those who have helped others to those who have hindered others (e.g. Jacob & Dupoux, 2008). Moreover, non-humans are also similarly sensitive to reputation in choosing partners: chimpanzees, for example, have been shown to recruit the best possible collaborators for cooperative tasks on the basis of prior interactions (Melis, Hare, & Tomasello, 2006). And in some species, mere observation of third-party interactions is also sufficient. For example, clients of the reef cleaner fish, *labroides dimidiatus*, observe cleaning sessions involving other fish, and spend more time in the vicinity of cleaners whom they have observed to be cooperative. Indeed, the cleaners themselves behave more cooperatively when they are being observed (Bshary & Grutter, 2006).

The most straightforward way in which prior knowledge from either source (2nd or 3rd party) could lead a human agent to treat a robot as a candidate for commitment is simply to tell the human agent that the robot is designed in such a way that it can be trusted to honor commitments and to expect its human interactants also to do so. They could be told, for example, that the robot had been designed to value the prospect of developing a trusting relationship with humans, and to cultivate a good reputation. Moreover, third-parties could inform people that they had found a particular robot or type of robot to be trustworthy and perhaps to expect trustworthiness from people. In this connection, it is interesting that Milinski and colleagues (2002) have shown that such gossip can profoundly influence group behavior in various contexts. Sommerfeld, Krambeck, Semmann, and Milinski (2007) have even observed the same phenomenon, using a trust game, when participants could have accessed the relevant information by direct observation. And Delgado and colleagues (2005) found that participants made more risky investment decisions with partners when they had been informed by a third-party that the partner had a good reputation.

Alternatively, simply by interacting with a robot, a human could build up an expectation that the robot were trustworthy. This could be achieved, for example, if the human agent could observe that the robot acts as committed on several occasions despite being offered some outside option which it were known to find attractive (e.g. rescuing the life of a human being instead of preserving its own existence). In addition, the robot could perhaps even exhibit reluctance to perform the action to which it had committed, only to, hesitatingly, go through with it under great visible effort.

9.2.2 *Emotions*

Emotions are clearly of central importance in commitment – emotional expressions can signal commitment (Michael, 2011), may be crucial in motivating agents to honor commitments (Frank, 1988), and potentially also in enabling individuals to identify and assess the level of their own and others' commitments (i.e. by anticipating the likely emotional outcomes of reneging, such as guilt, shame, anger, disgust, etc.).

If this suggestion is on the right track, then the anticipation of emotional consequences of actions (e.g. guilt aversion, Charness & Dufwenberg, 2006; Ahn, Kim, & Aggarwal, 2014) might serve as a heuristic for assessing the reliability of commitments.[1] For example, in a case in which it would be considered justified for

[1]This is consistent with the idea, recently articulated by Szigeti (2013), that moral emotions serve as heuristics for assessing the moral status of actions.

YOU to feel guilt or shame for neglecting to contribute to ME's goal, or for ME to express anger, ME can be confident that YOU will make the contribution and thus avoid unpleasant emotional consequences. Such an emotion-based heuristic may in fact serve as a proximal mechanism in connection with the maintenance of reputation. One consequence of this would be that individuals who do not experience moral emotions in a typical manner or do not understand them as others in their culture do, may also exhibit an anomalous understanding of commitments. In the context of development, this would imply that children's understanding of commitments should depend upon the development of their ability to anticipate moral emotions. Let us briefly consider some data that bears upon this conjecture.

First of all, the predominant view in developmental psychology is that children begin to exhibit pride and embarrassment around their second birthdays, showing public elation when performing well at difficult tasks, and blushing and hiding their faces when they do not do well at some task or other. It is noteworthy that this is around the time when they first pass the mirror test (Bischof-Köhler, 1991), given that these emotions depend upon a self-other distinction and an understanding of how one appears from the outside, i.e. to the gaze of other people. As Philippe Rochat (2008, 249) puts it:

> In such secondary or self-conscious emotions, children demonstrate unambiguously that what they hold as representation of themselves (i.e. self-knowledge) factors the view of others. They begin to have others in mind in how they construe who they are. With secondary emotions such as embarrassment, pride, or contempt, the child further demonstrates the triadic nature of self-knowledge, a knowledge that is coconstructed with others …

Furthermore, it is also worth noting that Rakoczy and colleagues (Rakoczy, 2008; Rakoczy & Schmidt, 2013; Schmidt, Rakoczy, & Tomasello, 2011) have provided evidence that children are sensitive to social norms by around 24 months, and even inclined to enforce them by protesting against violators.

But, of course, exhibiting or experiencing such emotions is different from understanding or anticipating them. And this is consistent with the finding that an understanding of complex moral emotions, such as guilt, pride and shame, continues to undergo fundamental development until at least around seven or eight years (Harris, 1989; Harris, Olthof, Terwogt, & Hardman, 1987; Nunner-Winkler & Sodian, 1988). Interestingly, children under this age rarely refer to such complex emotions in their speech (Ridgeway, Waters, & Kuczaj, 1985), and when presented with vignettes where an agent either succeeds or fails at some action with a moral significance according to their effort, their luck, or outside intervention, children younger than seven years are not proficient at inferring the resultant moral emotions, such as shame, guilt, pride and anger (Thompson, 1987; Thompson & Paris, 1981; Weiner, Graham, Stern, & Lawson, 1982). Barden, Zelko, Duncan, and Masters (1980) also reported that four-five year-olds predicted that a person would be "happy" if they had committed an immoral act but not been caught, whereas nine-ten-year-olds predicted that they would be scared or sad. When asked to predict their own emotions in such situations, the children exhibited the same pattern. If the anticipation of moral emotions serves as an important heuristic for the tracking of commitments, then we should expect that children under about nine years should evince difficulties in some cases – in particular in making explicit judgments about violations of commitments.

Given this general timetable for the development of moral emotions and the ability to understand and anticipate them, we should expect children to begin to act in accordance with commitments, and to protest when others fail to, around their second birthdays. But we should not expect them to reliably anticipate whether people are likely to honor commitments, or to make reliable judgments about commitment violations, until around nine years.

And indeed this pattern is strikingly confirmed by the existing data. First of all, Warneken and colleagues (2006) found that children as young as 18 months would protest when an experimenter with whom they were engaged in a simple joint action abruptly disengaged, thereby reneging on an implicit commitment to remain engaged until both parties to the joint action were satisfied that it had been completed. In a follow-up to this study, Gräfenhain and colleagues (2009) introduced a distinction between a condition in which the experimenter made an explicit commitment to the joint action and a condition in which she simply entered into the action without making any commitment. Their finding was that three-year-olds, but not two-year-olds, protested significantly more when a commitment had been violated than when there had been no commitment. Moreover, in experiment 2 of the same study, the tables were turned and the children presented with an enticing outside option which tempted *them* to abandon the joint action. The finding here was that the children were less likely to succumb to the temptation if a commitment had been made, and in cases in which they did succumb, they were more likely to "take leave", to look back at the experiment nervously, or to return after a brief absence. These findings appear to suggest that they understand something about commitments by around three years. Further support comes from another study by Hamann and colleagues (2012), in which one child received her part of a joint reward from a joint task before her partner received the other part, thus testing their understanding that they should remain engaged until both parties achieve their goal.

Note, however, that the children in these studies did not have to make judgments about commitments or anticipate future behavior. Indeed, both in experiment 2 of the Gräfenhain study and in the Hamann study, the children frequently did first abandon the joint action and then return to it (in the Hamann study, the vast majority of the three-year-olds only returned to the joint action when the other child protested), demonstrating that they had failed to anticipate the negative emotional consequences of doing so. Presumably, they learn this gradually in the coming years and become much more proficient at making decisions that will not lead to negative emotional outcomes. This may take quite some time, however, as is demonstrated by two other studies probing children's understanding of commitments. First, consider a study conducted by Mant and Perner (1988), in which children were presented with vignettes in which two children on their way home from school, Peter and Fiona, discuss whether to meet up and go swimming later on. In one condition, they make a joint commitment to meet at a certain time and place, but Peter decides not to go after all, and Fiona winds up alone and disappointed. In the other condition, they do not make a joint commitment, because Fiona believes that her parents will not let her. She is then surprised that her parents do give her permission, and she goes to the swimming hole to meet Peter. In this condition, too, however, Peter decides not to go after all, so again Fiona winds up alone and disappointed. The children in the study, ranging from five to ten years of age, were then asked to rate how naughty each character was. The finding was that only the oldest children (with a mean age of 9.5), and not the younger children, differentiated between the commitment condition and the no-commitment condition in rating Peter's level of naughtiness. This may sound surprisingly late, but it is consistent with the findings of Astington (1988), who reported that children under 9 did not understand that one can only promise to bring about events over which one has some control, and that children as old as 11–13 judged that a speaker had not made a promise at all in cases in which the promise was unfulfilled.

What about other species? Comparative research supports the existence of social norms governing play behavior in some species. Dogs and other canids, for example, commonly initiate and punctuate play sessions with stock play signals, such as bowing forward, thereby generating an expectation on the part of potential playmates that they will not bite or otherwise cause harm (Bekoff, 1995). It has also been observed that individual coyotes who bite after having given such play signals tend to be excluded from future play, suggesting that their playmates are sensitive to their having violated a minimal commitment to play nicely, and accordingly inclined to impose sanctions (Bekoff, 2002).

Of course, this does not demonstrate that canids anticipate the negative social consequences of violating these commitments, nor that they refrain from doing so because of the likely consequences. But such a capacity would be very useful to any agents who possess it, since it would make it possible to anticipate one's own and others' emotional responses to various actions, as well as the effects upon one's own and others' reputations, and to use this information in assessing the likelihood of others' making expected contributions and the consequences of one's own doing so or not doing so.

One challenge to be confronted by roboticists seeking to implement emotions components of commitments in robots is that robots have different bodies from humans and may therefore be unable to express emotions in a way that elicits the same type of response from human interactants as human expressions of emotions. Indeed, in view of the well-known uncanny valley effect (Mori, 1970), it may be counterproductive to design robots such that they express emotions in a way that too closely parallels human emotional expressiveness.

In order to address this challenge, it would be useful to identify particular forms of expression that work well in humans. For example, perhaps facial expressions of emotion would not be the most feasible (in view of the uncanny valley effect), but bodily posture or tone of voice might be more effective. Alternatively, people could simply be told that a particular robot or type of robot has emotions, such as guilt, fear or disgust, which the robot might even be able to express in a peculiar non-human way, and that these emotions motivate it to act in accordance with commitments and to expect others to do so. It is interesting in this context to note that humans do tend to ascribe feelings to robots with a non-human appearance, apparently on the basis of contextual cues: for example, Rosenthal-von der Pütten, Krämer, Hoffmann, Sobieraj, and Eimler (2013) reported that human participants reacted empathically (i.e. rejoicing or commiserating) to images of a dinosaur-robot either being treated in a friendly way or being tortured.

Finally, irrespective of robots' capacity to express emotions in a manner that humans find compelling, it could be important to develop their capacity to monitor and respond sensitively to humans' emotions. It may be that people would be more inclined to trust, and to feel committed toward, robots who had exhibited a concern for their emotions.

9.2.3 Bodily Cues

There are several interesting studies showing the importance of bodily cues, such as eye contact and characteristically cooperative movements dynamics, in human-human interactions. For example, Wyman, Rakoczy, and Tomasello (2013) devised a stag hunt game in which four-year-old children could choose to attempt to attain a large prize (i.e. the stag, in this case represented by cool stickers) or a small prize (i.e. the hare, in this case represented by less cool stickers). In order to attain the large prize, they had to operate a large lever which could only be operated if the other player also opted to go for the larger prize, whereas they could obtain the smaller prize by operating a smaller lever by themselves. The interesting finding was that the children were far more likely to go for the large prize if there had been eye contact with the other player, i.e. signaling an implicit commitment to coordinate efforts in order to obtain the larger prize. The notion that eye contact generates an implicit commitment to be cooperative gains further support from and interesting study by Bayliss and Tipper (2006). In a gaze-cueing paradigm, they presented participants with faces gazing either to the right or to the left, and participants that either reliably cued, or reliably failed to cue, the location of an object which appeared shortly thereafter, and which participants had the task of detecting. What they found was that participants were equally likely to follow the gaze of the seen faces, regardless of how helpful this was, but that they tended to rate the unhelpful faces as less trustworthy than the helpful faces – as thought the unhelpful faces had violated an implicit commitment.

An obvious challenge here is that robots' specific bodily forms may make it difficult for them to give such signals as easily as humans. However, it is worth noting that human participants respond to cartoon faces (Friesen & Kingstone, 2003; Böckler, Sebanz, & Knoblich, 2011) much in the same way as they respond to human faces in gaze-cueing paradigms, i.e. they tend to follow gaze direction. Indeed, Johnson, Slaughter, and Carey (1998) found that even young children tend to follow the gaze direction of robots as long as those robots act in a way that is contingent upon their actions, and thus appear to be contingently responsive to them. Intriguingly, in one experiment, the robots did not even have faces but merely front and rear sides. Hence, it is possible to design robots that can engage in eye contact even without having faces, and certainly without having human-like faces.

In a very different experimental setting, Pezzulo and Dindo (2011) found that participants tended to sacrifice efficiency of movement in order to make their movements more easily and quickly predictable for their partners (see also Vesper et al., 2010). Insofar as this type of signaling constitutes an investment in the joint action, and demonstrates a willingness to coordinate with one's partner, it may serve to reinforce the expectation that the agent doing the signaling will remain engaged until the goal is reached. If so, it is plausible that signaling could also have such effects in human-robot interactions. After all, there is no need to have the same bodily shape as a human in order to adapt one's movements to make them more predictable. Work by Lee, Knox, Wormwood, Breazeal, and Desteno (2013) has shown that it is possible to identify bodily cues that correlate with trust in dyadic interactions, and to design robots to exhibit and to identify such cues.

9.2.4 Conversational Expertise

A further element that might lead a human agent to the idea that a robot can engage in a commitment is the robot's expertise in conversation. According to an accredited paradigm in the philosophy of language (cf. Grice, 1989), conversation is an activity that requires cooperation. The idea here is that one's contribution to a conversation has to be appropriate to the accepted purpose and direction of the conversation (call this the "Cooperative Principle"). More specifically, the contribution has to be true, relevant, perspicuous, understandable, etc. That is, humans engaging in communicative intercourse expect that their interactants will contribute to the conversation based on a set of implicit rules such that by entering a conversation one also enters into certain commitments to follow those rules.

This grounds the possibility of so-called conversational implicatures, i.e., a kind of inference that a hearer can draw based upon the conventional meaning of an utterance and upon the context in which the utterance has been made. For instance, if Alessandro says to John "I am out of petrol" and John replies "Turn left," then – since Alessandro assumes that John is committed to act according to the Cooperative Principle and hence to deliver relevant contributions to the conversation–Alessandro has reason to infer that John knows that there is a gas station to the left. Accordingly, if a robot could demonstrate the ability to communicate not only based on conventional meanings (i.e., on the semantic meanings of sentences), but also in a manner that were sensitive to a human hearer's ability to process conversational implicatures, the human hearer might come to the impression that the robot entered into certain commitments when it engaged in conversation.

Two interrelated conclusions can be drawn from these considerations. The first is that, by engaging in a conversation with a robot, a human agent might come to the conviction that the robot shared what recently has been qualified as the "conceptual common ground" of communication, i.e., everything that the speakers both know (and know that they both know, Clark, 1996). More specifically, a robot's sensitivity to conversational implicature may lead a human agent to form the following beliefs (cf. Tomasello 2008):

(i) that the robot can engage in "joint attention" with the human (as in cases where I say "*this* is amazing" and John understands I am talking about a glass of *Amarone* because we both are seeing this glass of red wine *together*);
(ii) that the robot has access to the common ground created by top-down processes – e.g., John and I are pursuing a goal together and so we both know what is relevant to both of us right now, given that goal;
(iii) that the robot has access to common cultural knowledge – i.e. to knowledge that is not explicitly acknowledged, but simply assumed on the basis of cultural markers.

Point (iii), in particular, may lead a human interactant to expect a robot to understand that certain kinds of activity (communication, e.g.) require participants to follow a set of rules and, more precisely, to follow rules that have not been explicitly formulated, publicly laid out and negotiated prior to the activity. The fact that the robot masters those rules would show to the human agent that it draws upon the norms and, in certain cases, the values (think of values like politeness, eloquence, etc.) of the community to which the human agent herself belongs. This could also make it possible for human agents to sense that the robot in question understood the norms governing the degree of commitment that is appropriate, and the circumstances under which it is considered acceptable within the community to renege on commitments, i.e. without the other party becoming angry or distrustful.

There is a second, and for the purpose of this paper probably even more important, issue related to robots' pragmatic expertise in communication. The possibility to engage in a conversation with a robot in the way just sketched above could be taken by a human agent to be a signal that the robot has an understanding of what it means to have a commitment to following the implicit rules of a conversation. Indeed, if the Cooperative Principle holds, then all participants are expected to observe it, i.e., they are committed to following it. The crucial point is not so much that, as we stated above, these rules generally are implicit (although the ability to understand implicit rules is certainly impressive in itself) but, rather, that having a commitment towards the (implicit or explicit) rules governing an activity demonstrates an interest in sustaining rather than merely following the rules regulating the activity at issue. It seems that, if a robot can communicate by exploiting a human hearer's capacity to process conversational implicatures, then the hearer is provided with some evidence that the robot is honoring a commitment to follow the "rules of the (conversational) game" as such, and not only those rules that are relevant for specific conversational situations. A similar mechanism seems to be at work when it comes to shared cooperative activities, and especially to the commitment to mutual supportiveness into which parties to a joint action commonly enter. And it is to this issue that we now turn our attention.

9.2.5 Meshing Sub-plans and Mutual Supportiveness

As philosophers have observed (Bratman, 2014; Gilbert, 1989; Searle, 2002), a commitment towards a shared goal is characteristic of joint actions among humans. However, this is not yet sufficient for a so-called "shared cooperative activity (SCA)" to occur. Two individuals could both be committed to painting the house together, and yet one of them may want to do it in red, the other in blue. That is, SCA also requires the participants to carry out complementary, or meshing, subplans, and to be able and willing to flexibly adjust their subplans in order to bring about goals to which they are committed. This means that both agents are required to coordinate with each other: their intention to reach a shared goal is formed in accordance with and because of the meshing subplans (Salice, 2015). It has been noted, though, that commitment to a shared goal and mutual responsiveness, although necessary and sufficient for defining a *jointly intentional activity*, are not yet apt to define SCA. Bratman (1992) famously presents the case of two singers who coordinate in a duet, but who are not willing to help each other (and indeed would rejoice) if their counterpart failed. What is still lacking in the picture of full-blown SCA is that the cooperators are typically also willing, and indeed take each to be obliged, to offer mutual support where necessary. Such a commitment to mutual support may underpin a "mutual assumption of helpfulness," which is central to the notion of shared intentionality (Tomasello, 2008, 2014).

How can all this be transferred to the case of human-robot cooperation? It seems that, given that the technical challenges can in principle all be met, robots should not have any particular difficulties with mutual responsiveness. After all, it seems plausible to assume that a robot can coordinate with other (human or non-human) agents, provided that the task is well-defined. Some additional challenge could be presented in designing robots to be flexible enough to adapt their subplans in order to mesh with a human who, for example, was struggling with her own subplan. But, again, the benefit could be well worthwhile, insofar as a high level of engagement can be demonstrated by a tendency to monitor a partner's actions, and to flexibly adjust action plans to complement one's partner. Moreover, the implementation of a capacity to flexibly adjust subplans on the fly should be eminently feasible, as is illustrated, for example, by a recent project by Clodic et al. (2006). Clodic and colleagues designed an interactive robot guide, "Rackham", who made explicit agreements with museum visitors to guide them through an exhibition, and who was able to monitor whether his "clients" were following him, to wait for them, to adjust his pace to theirs, etc.

But, to return to the key question: when might a human agent come to sense that a robot is committed to the shared goal? In addition to all that has been said above, it could also be important for the robot to exhibit a readiness for mutual support – *even in contexts where no shared cooperative action is planned.* If the robot succeeds in creating the impression that it, in general, is supportive in a spontaneous and, to a certain extent, even unselfish way, and hence if the human agent has reasons to consider her assumption of helpfulness satisfied, then she might also expect the robot to honor its commitment towards a specific goal. Indeed, assuming that the robot is able to represent others' goals and the impediments that one could encounter in reaching one's goals, it could be programmed in such a way that in those salient situations in which help actually is needed, it also provides the needed help. As a consequence, if a robot appears to be willing and able to honor a general commitment to help people in need, then a human agent may very well expect it to be able to commit itself to a more specific goal – like those that define a SCA.

9.3 Joint Commitment and Group Identification

So far, our discussion has addressed the possibility of using the minimal framework to implement a sense of commitment in general in human-robot interaction. In this section, we will apply the minimal approach to the phenomenon of *joint* commitment. On the minimal approach, joint commitment can be analyzed (in a minimal sense) as the mere conjunction of two agents' sense that they are both committed, i.e. where two (or more) agents are motivated to contribute to a goal in part because the other is too, and where each expects this of the other. However, it is interesting to note that other, richer, analyses of joint commitment have also been offered. For example, Margaret Gilbert (1989, 2006, 2014) has emphasized that joint commitments involve the genesis of a group, i.e. a plurality of individuals that are glued together, as it were, in the efforts to reach a shared goal. Gilbert in fact takes the concept of joint commitment to be as so fundamental as to define the notion of a group. Thus, in her view, if certain individuals enter into a joint commitment, they can be said to be a group (or a "plural subject"). Some other philosophers conceptualize the relation the other way around: it is because individuals form a group that they are able to enter into joint commitments (e.g. Schmid, 2014; Tollefsen, 2002; Tuomela, 1992).

Whatever explanation one prefers, this richer understanding of joint commitment could be said to be intrinsically social in the sense that the author of a joint commitment is a "plural subject." Thus, joint commitments can ground group membership: either because (as in Gilbert's account) entering a joint commitment qualifies a subject as a member of the group, or because (as in the competing account) reneging on joint commitments may lead to group ostracism.

We do not take any stance on the question whether joint commitments ground groups or vice versa. Instead, we would like to emphasize that, whichever way one conceptualizes this relation, it is clear that group identity is intrinsically related to joint commitment. We take this to be an important point insofar it raises the following novel question: assuming that entering joint commitments either necessarily leads to group formation or it presupposes group formation, to what extent would humans be willing to form groups with robots?

In approaching this question, it will be helpful to briefly draw upon social identity theory, according to which individuals do not only have a personal identity but also a *social* identity – in fact, they typically have *several* social identities. The general idea is that individuals have a strong tendency (perhaps even motivated by a "need to belong", cf. Baumeister & Leary, 1995) to self-categorize by means of features or properties that they perceive sharing with other individuals and to which they assign salience – when this happens, they "group-identify." Put differently, they establish a social identity insofar as they think of themselves as being a member of the group defined by the feature(s) at stake. Accordingly, social identity or the social self can be intended as the representation one has of oneself *qua* member of a *group* (Brewer, 1991). The flipside of this process is that group-identification goes hand in hand with the definition of an out-group, which is delimited by the class of individuals that do not share the property characterizing in-group membership.

On the one hand, these insights from social identity theory could be taken to represent a challenge to the notion that humans may form groups with robots (and therefore enter joint commitments with them). This is because a human's awareness that her/his interactant is a robot may trigger the assignment of the robot to an out-group, i.e., to the group of non-human agents. This could counteract, or indeed block, group-identification and thus also the predisposition to enter into joint commitments with robots.

On the other hand, we think that this way of formulating the challenge is oversimplified. First of all, a human agent's belief that her or his robot interactant is a robot and not a human is not likely to explicit most of the time. Instead, it is likely to have the character of more or less vague background knowledge that is susceptible to variation and influence. Secondly, as we were at pains to show in Sect. 9.2, people often treat robots (and other artifacts) as though they were conscious agents with a free will even despite explicitly denying that this is the case. Thus, there is no reason in principle why humans should not be more or less willing to form groups with robots on the basis of the same types of cue that play a role in human-human group identification. True, some such cues are likely to be less available, or even completely unavailable (e.g. having the same parents). But for some other cues, such as shirt color, objective membership in organizations or states, etc. there may be little or no difference between human-human and human-robot interaction.

In sum, it may be useful for roboticists moving forward to try to identify what cues work well in leading humans to group-identify with robots and thus to be increase their willingness to enter into joint commitments with robots.

9.4 Conclusions

In this paper we have discussed the notion of commitment in its relation to human-robot interaction. We tackled some of the central challenges that social roboticists might face in designing robots with which human agents would be inclined to engage in commitments, by focusing especially on joint commitments and on the link that this type of commitment has to group identification.

There are deeper philosophical questions which our pragmatic approach has not addressed. In particular, we did not address the question as to whether a human agent would be right in thinking that a robot interactant had the capacity to honor commitments and to demand the same in return. In other words, we left the question open about whether and under what conditions robots can really engage in commitments. Although answering this question would require a larger investigation of deep conceptual issues, we are convinced that, whatever its answer, the potential is immense for robots with whom human agents would be willing to engage in joint activities and their attendant commitments.[2]

Acknowledgements We are grateful to Raul Hakli, Johanna Seibt, Anna Strasser and Chris Frith for helpful comments on earlier versions of this paper. John Michael's work on this paper was made possible by a Marie Curie Intra-European Fellowship PIEF-GA-2012-331140) and by ERC Starting Grant Agreement no. 679092, The Sense of Commitment).

References

Ahn, H.-K., Kim, H. J., & Aggarwal, P. (2014). Helping fellow beings: Anthropomorphized social causes and the role of anticipatory guilt. *Psychological Science, 25*, 224.

Astington, J. W. (1988). Children's understanding of the speech act of promising. *Journal of Child Language, 15*, 157–173.

Axelrod, R. (1984). *The evolution of cooperation*. New York: Basic Books.

Barden, R. C., Zelko, F. A., Duncan, S. W., & Masters, J. C. (1980). Children's consensual knowledge about the experiential determinants of emotion. *Journal of Personality and Social Psychology, 39*(5), 968.

Baumeister, R. F., & Leary, M. R. (1995). The need to belong: Desire for interpersonal attachments as a fundamental human motivation. *Psychological Bulletin, 117*(3), 497–529.

Bayliss, A. P., & Tipper, S. P. (2006). Predictive gaze cues and personality judgments: Should eye trust you? *Psychological Science, 17*, 514–520.

Bekoff, M. (1995). Play signals as punctuation: The structure of social play in Canids. *Behaviour, 132*, 419–429.

Bekoff, M. (2002). Social play behavior: Cooperation, trust, fairness and the evolution of morality. *Journal of Consciousness Studies, 8*, 81–90.

Bischof-Köhler, D. (1991). The development of empathy in infants. In M. E. Lamb & H. Keller (Eds.), *Infant development: Perspectives from German-speaking countries* (pp. 245–273). Hillsdale, NJ: Lawrence Erlbaum Associates.

Böckler, A., Sebanz, N., & Knoblich, G. (2011). Observing shared attention modulates gaze following. *Cognition, 120*, 292–298.

Bratman, M. E. (1987). *Intention, plans, and practical reason*. Cambridge, MA: Harvard University Press.

Bratman, M. E. (1992). Shared cooperative activity. *The Philosophical Review, 101*, 327–341.

Bratman, M. E. (2014). *Shared agency: A planning theory of acting together*. Oxford: Oxford University Press.

Breazeal, C. (2004). Social interactions in HRI: The robot view. *IEEE Transactions on Systems, Man, and Cybernetics, Part C (Applications and Reviews), 34*(2), 181–186.

[2]Although we have emphasized some advantages of implementing minimal commitments in HRI at the beginning of this paper, its potential risks should not be underestimated. Although the ethical aspects of forms of HRI based upon the minimal approach to commitment are crucial in many respects, their investigation would exceed the scope of this paper.

Brewer, M. B. (1991). The social self: On being the same and different at the same time. *Personality and Social Psychology Bulletin, 17*, 475–482.

Bshary, R., & Grutter, A. (2006). Image scoring and cooperation in a cleaner fish mutualism. *Nature, 441*, 975–978.

Charness, G., & Dufwenberg, M. (2006). Promises and partnership. *Econometrica, 74*(6), 1579–1601.

Clark, H. H. (1996). *Using language*. Cambridge: Cambridge University Press.

Clark, H. H.(2006). Social actions, social commitments. In N. J. Enfield & S. C. Levinson (Eds.), *Roots of human sociality: Culture, cognition, and interaction* (pp. 126–150). Oxford/New York: Berg Publishers.

Clodic, A., Fleury S., Alami, R., Chatila, R., Bailly, G., Brethes, L., ... others (2006). Rackham: An interactive robot-guide. In *ROMAN 2006-the 15th IEEE International Symposium on Robot and Human Interactive Communication*, Hatfield (pp. 502–509).

Cohen, P. R., & Levesque, H. J. (1990). Intention is choice with commitment. *Artificial Intelligence, 42*, 213–261.

Delgado, M. R., Frank, R. H., & Phelps, E. A. (2005). Perceptions of moral character modulate the neural systems of reward during the trust game. *Nature Neuroscience, 8*(11), 1611–1618.

Frank, R. H. (1988). *Passions within reason: The strategic role of the emotions*. New York: WW Norton & Co.

Friesen, C. K., & Kingstone, A. (2003). Covert and overt orienting to gaze direction cues and the effects of fixation offset. *NeuroReport, 14*(3), 489–493.

Gilbert, M. (1989). *On social facts*. Princeton, NJ: Princeton University Press.

Gilbert, M. (2006). Rationality in collective action. *Philosophy of the Social Sciences, 36*(1), 3–17.

Gilbert, M. (2014). *Joint commitment: How we make the social world*. Oxford: Oxford University Press.

Gräfenhain, M., Behne, T., Carpenter, M., & Tomasello, M. (2009). Young children's understanding of joint commitments. *Developmental Psychology, 45*(5), 1430–1443.

Grice, P. (1989). *Studies in the way of words*. Cambridge: Harvard University Press.

Hamann, K., Warneken, F., & Tomasello, M. (2012). Children's developing commitments to joint goals. *Child Development, 83*(1), 137–145.

Harris, P. L. (1989). *Children and emotion: The development of psychological understanding*. Oxford: Blackwell.

Harris, P. L., Olthof, T., Terwogt, M. M., & Hardman, C. E. (1987). Children's knowledge of the situations that provoke emotion. *International Journal of Behavioral Development, 10*(3), 319–343.

Hollander, J., & Wijnberg, J. (2009). *What is MindMentor?* Retrieved from http://mindmentor.com/index.php?pid=learnmore

Jacob, P., & Dupoux, E. (2008). Developmental psychology: A precursor of moral judgment in human infants? *Current Biology, 18*, R216–R218.

Johnson, S., Slaughter, V., & Carey, S. (1998). Whose gaze will infants follow? The elicitation of gaze-following in 12-month olds. *Developmental Science, 1*(2), 233–238.

Kahn, P. H., Kanda, T., Ishiguro, H., Gill, B. T., Shen, S., Gary, H. E., & Ruckert, J. H. (2015) Will people keep the secret of a humanoid robot?: Psychological intimacy in HRI. In *Proceedings of the Tenth Annual ACM/IEEE International Conference on Human-Robot Interaction*, Portland (pp. 173–180).

Kahneman, D., & Tversky, A. (1979). Prospect theory: An analysis of decision under risk. *Econometrica, 47*(2), 263–292.

Lee, J., Knox, W., Wormwood, J., Breazeal, C., & Desteno, D. (2013). Computationally modeling interpersonal trust. *Frontiers in Psychology, 4*(893), 1–14.

Mant, C. M., & Perner, J. (1988). The child's understanding of commitment. *Developmental Psychology, 24*(3), 343.

Melis, A., Hare, B., & Tomasello, M. (2006). Chimpanzees recruit the best collaborators. *Science, 311*, 1297–1300.

Michael, J. (2011). Shared emotions and joint action. *Review of Philosophy and Psychology, 2*(2), 355–373.

Michael, J., Sebanz, N., & Knoblich, G. (2016). The sense of commitment: A minimal approach. *Frontiers in Psychology, 6*, 1968.

Milinski, M., Semmann, D., & Krambeck, H.-J. (2002). Reputation helps solve the tragedy of the commons. *Nature, 415*(6870), 424–426.

Mordoch, E., Osterreicher, A., Guse, L., Roger K., & Thompson, G. (2013). Use of social commitment robots in the care of elderly people with dementia: A literature review. *Maturitas, 74*, 14–20.

Mori, M. (1970). Bukimi no tani. *Energy, 7*(4), 33–35.

Nunner-Winkler, G., & Sodian, B. (1988). Children's understanding of moral emotions. *Child Development, 59*, 1323–1338.

Pacherie, E. (2013). Intentional joint agency: Shared intention lite. *Synthese, 190*, 1817–1839.

Pezzulo, G., & Dindo, H. (2011). What should I do next? Using shared representations to solve interaction problems. *Experimental Brain Research, 211*, 613–630.

Rakoczy, H. (2008). Taking fiction seriously: Young children understand the normative structure of joint pretend games. *Developmental Psychology, 44*, 1195–1201.

Rakoczy H., & Schmidt, M. (2013). The early ontogeny of social norms. *Child Development Perspectives, 7*, 17–21.

Ridgeway, D., Waters, E., & Kuczaj, S. A. (1985). Acquisition of emotion-descriptive language: Receptive and productive vocabulary norms for ages 18 months to 6 years. *Developmental Psychology, 21*(5), 901.

Rochat, P. (2008). "Know thyself!"... but what, how and why? In F. Sani (Ed.), *Individual and collective self-continuity: Psychological perspectives* (pp. 243–251). Mahwah, NJ: Lawrence Erlbaum Publishers.

Rosenthal-von der Pütten, A. M., Krämer, N. C., Hoffmann, L., Sobieraj, S., & Eimler S. C. (2013). An experimental study on emotional reactions towards a robot. *International Journal of Social Robotics, 5*(1), 17–34.

Salice, A. (2015). There are no primitive we-intentions. *Review of Philosophy and Psychology, 6*(4), 695–715.

Schmid, H. B. (2014). Plural self-awareness. *Phenomenology and the Cognitive Sciences, 13*, 7–24.

Schmidt, M., Rakoczy, H., & Tomasello, M. (2011). Young children attribute normativity to novel actions without pedagogy or normative language. *Developmental Science, 14*, 530–539.

Searle, J. R. (2002). Collective intentions and actions. In *Consciousness and language* (pp. 90–105). Cambridge: Cambridge University Press.

Searle, J. R. (2010). *Making the social world: The structure of human civilization*. Oxford: Oxford University Press.

Skyrms, B. (2004). *The stag hunt and the evolution of social structure*. Cambridge: Cambridge University Press.

Sommerfeld, R. D., Krambeck, H.-J., Semmann, D., & Milinski, M. (2007). Gossip as an alternative for direct observation in games of indirect reciprocity. *Proceedings of the National Academy of Sciences, 104*(44), 17435–17440.

Szigeti, A. (2013). No need to get emotional? Emotions and heuristics. *Ethical Theory and Moral Practice, 16*(4), 845–862.

Thompson, R. A. (1987). Development of children's inferences of the emotions of others. *Developmental Psychology, 23*(1), 124–131.

Thompson, R. A., & Paris, S. G. (1981). *Children's inferences about the emotions of others*. Paper presented at the biannual meeting of the Society for Research in Child Development, Boston.

Tollefsen, D. P. (2002). Collective intentionality and the social sciences. *Philosophy of the Social Sciences, 32*(1), 25–50.

Tomasello, M. (2008). *Origins of human communication*. Cambridge: MIT Press.

Tomasello, M. (2014). *A natural history of human thinking*. Cambridge, MA: Harvard University Press.

Tuomela, R. (1992). Group beliefs. *Synthese, 91*(3), 285–318.

Vesper, C., Butterfill, S., Knoblich, G., & Sebanz, N. (2010). A minimal architecture for joint action. *Neural Networks, 23*(8–9), 998–1003.

Warneken, F., Chen, F., & Tomasello, M. (2006). Cooperative activities in young children and chimpanzees. *Child Development, 77*(3), 640–663.

Weiner, B., Graham, S., Stern, P., & Lawson, M. E. (1982). Using affective cues to infer causal thoughts. *Developmental Psychology, 18*(2), 278.

Wyman, E., Rakoczy H., & Tomasello, M. (2013). Non-verbal communication enables coordination with others in a children's 'stag hunt' game. *European Journal of Developmental Psychology, 10*(5), 597–610.

Chapter 10
The Problem of Understanding Social Norms and What it Would Take for Robots to Solve it

Johannes L. Brandl and Frank Esken

Abstract In this paper, we argue that there is no evidence in sight warranting the conclusion that robots are social agents in some strong sense that requires an understanding of social norms. In support of this skepticism, we first consider an argument to the effect that a basic sensitivity to norms requires no mindreading abilities and may therefore also be found in non-human animals. In rebutting this view, we rely on arguments based on Searle's theory of institutional facts and Tomasello's theory of collective intentionality. We, then, extend these arguments to the case of robots and argue that robots' social behaviour does not reach the level at which an understanding of norms becomes crucial.

Keywords Norm understanding • Adaptation • Regularities • Norms • Sensitivity • Animal behavior • Child development • Social robots

10.1 Introduction

Robots can do many things that are best described in social terms: they engage with others (i.e. with other robots or with natural beings like us) in various forms of joint action, often cooperating with others in solving problems, and they can guide and teach us humans much like a human instructor might do. Should we therefore treat robots as genuinely social agents? What matters here most is the question whether robots can understand social norms. In addressing this question, we propose to take an indirect route. We will first consider an argument for the claim that some non-human animals – notably chimpanzees – possess a primitive sense of social normativity (Sect. 10.2). We then confront this claim with two prominent objections: one objection is based on Searle's theory of social facts, the other on Tomasello's claim that only humans possess the kind of collective

---

This paper is thoroughly co-authored. The order of the authors is purely alphabetical.

J.L. Brandl (✉) • F. Esken
Institute of Philosophy, University of Salzburg, Salzburg, Austria
e-mail: johannes.brandl@sbg.ac.at; frankesken@mac.com

R. Hakli, J. Seibt (eds.), *Sociality and Normativity for Robots*, Studies in the Philosophy of Sociality 9, DOI 10.1007/978-3-319-53133-5_10

intentionality that is necessary for understanding social norms (Sect. 10.3). We then want to show why these considerations do not exhaust the problem of understanding social norms. As recent work on children's protest behavior has revealed, doubts remain whether the conditions specified by Searle and Tomasello are sufficient for a genuinely normative understanding of conventional rules (Sect. 10.4). These discussions lead us to distinguish three stages of social intelligence and argue that genuine understanding of social norms occurs only on the third stage (Sect. 10.5). Finally, we suggest that even on a generous interpretation of the available evidence about robots' social behavior, they reach only something in between the first and the second stage (Sect. 10.6).

10.2 The Case of Animals

Not only humans, but many non-human animals are highly social creatures. From this fact alone, one might infer that understanding social norms is not a human-specific capacity. However, ascribing such an understanding to animals has turned out to be highly controversial. A prominent recent advocate of the claim that non-human animals have a basic sense of social normativity has been Kristin Andrews (see Andrews, 2009, 2013). We want to take her view as a starting point for our discussion.

The case that Andrews builds for ascribing a primitive sense of social norms to animals is partly conceptual and partly empirical. On the one hand, Andrews relies on a conception of primitive normativity that takes its inspiration from Kant and Wittgenstein. According to this conception, we possess a primitive sense of normativity that requires neither an understanding of reasons for action nor an understanding of explicitly stated rules (see Ginsborg, 2011). For Andrews, this suggests that one can have such a primitive sense of normativity also without yet possessing a theory of mind, i.e. without possessing the ability to attribute to oneself or to others explicit knowledge that one is following a rule or a norm. This is what opens up the door for ascribing a sense of normativity to animals that may qualify as an implicit form of understanding norms, in contrast to a reflective, explicit understanding that is bound to language and to mindreading abilities (cf. Andrews, 2009, 2013).

Andrews appeals here to a venerable tradition in philosophy in support of her distinction between implicit and explicit understanding of norms. However, it is not philosophical reasoning *per se* that does the real work for her. Such a distinction remains empty as long as there are no criteria given for determining what kinds of behavior count as an implicit understanding of norms, with the additional requirement that such behavior cannot be explained in non-normative terms. Let us therefore turn now to the empirical evidence for ascribing such an implicit understanding of normativity to non-human animals. Andrews thinks that there is ample evidence, notably in the social behavior of chimpanzees. Here is one telling example that she cites as evidence for a norm violation:

> Chimpanzee infants are given deferential treatment by other community members. Newborn chimpanzees are extremely interesting to other community members, and adults will watch intently but not try to approach the new member of their group. Juveniles and older infants, however, act on their interest by trying to approach or touch the new infant, which leads the mother to act hostilely or defensively; these young chimpanzees quickly learn not to approach newborn infants. Once infants are old enough to venture away from their mother, adults are extremely tolerant of infants climbing over them and even stealing their food or tools. Adults are also reported to self-handicap when playing with infants. A violation of the norms regarding infants, such as infanticide or other other aggressive acts towards infants, often results in an uproar of vocal protest from adult females, and can also cause third party intervention. (Andrews, 2013, 194)

We do not deny that examples like this create a strong impression that chimpanzees possess something like a moral sense or at least understand something about social norms. But the mere fact that chimpanzees show emotions that have moral relevance from our point of view, does not necessarily mean that they have this relevance also from the apes' point of view. Andrews recognizes that there is still a line one can draw here. She therefore speaks cautiously only of "proto-social norms" manifest in behavioral regularities that "when violated, elicit bystanders reactions, including punishment and the expression of moral emotions" (Andrews, 2013, 194; see also Von Rohr, Burkart, & van Schaik, 2011). Andrews argues that apes possess an implicit understanding of social norms and moral constraints in the sense that they track behavioral regularities in their society, which from a human point of view could be called social norms. However, she is less cautious when she claims that this tracking, together with a motivation to follow the regularities they observe in their conspecifics, results "in a moral status that is not unlike that of many humans." (Andrews, 2013, 194).

So the question remains: Are these proto-social norms really norms or only proto-norms, i.e. social regulations that precede the formation of genuinely social norms?

10.3 Two Objections

In this section, we want to consider two lines of argument in support of a skeptical response to Andrews' view. One of them is based on the work of John Searle on social reality, the other can be found in Michael Tomasello's work on the origins of human thinking. Of course, we cannot provide here a comprehensive summary of their complex theories. We must confine ourselves to a raw sketch of the positions advocated by Searle and Tomasello.

A basic point in Searle's theory is that human society is a highly "institutionalized" construct. As we understand his view, this makes human society fundamentally different from even the most sophisticated animal societies. Although this may sound like a strong metaphysical claim, Searle takes it as an obvious fact that does not commit him to an unbridgeable gap between our species and the animal kingdom. When he contrasts humans and other animals, Searle says, "the point is not to make a plea for the superiority of our our species" (Searle, 2010,

p. 7 fn.). Nothing in his theory of social facts rules out the possibility that some non-human creature might one day show all the features of a human society, including “income tax, presidential elections, divorce courts, and other institutional facts” (Searle, 2010, p. 7 fn.). However, it seems hard to deny that humans are the only ones to have created such facts to date. Therefore, Searle does not hesitate to describe the phenomena he is interested in, as “distinctive human phenomena” (Searle, 2010, p. 7 fn.).

Of course, it is no accident that as far as we know humans are the only ones to have created institutional facts. The reason for this can be found in the different functions that humans impose on objects, in contrast to the functions imposed by animals. Searle writes:

> [Non-human] animals can impose functions on natural phenomena. Consider, for example, the primates that use a stick as a tool to get bananas that are out of reach. And some primates have even developed traditions of agentive functions that are transmitted from one generation to the next. Thus, most famously Imo, a Japanese macaque, used water to get the sand off her potatoes [...]. Thanks to Imo, today potato-washing in salt water is an established tradition which infants learn from their mother as a natural adjunct of eating potatoes. [...]. But the truly radical break with other forms of life comes when humans, through collective intentionality, impose functions on phenomena where the function cannot be achieved solely in virtue of physics and chemistry, but requires continued human cooperation in the specific forms of recognition, acceptance, and acknowledgment of a new status to which a function is assigned. (Searle, 1995, p. 40)

How should we take this claim about a “radical break” in forms of life? One way to understand Searle’s view here would go like this: Social facts are facts that result from the assignment of functions, and such assignments can be grounded in forms of collective intentionality that are part of animal life. Therefore, Searle is open to the view that social norms are not necessarily tied to institutional facts. If Imo learns from his mother how to properly wash the potatoes, he has learned a social norm without having learned an institutional fact.[1]

On this interpretation, the “radical break” would not seem to be very radical at all. First of all, it has not been shown that there can be social norms only in a society with an “institutionalized” structure as Searle describes it. For instance, it is a social norm in our society to cover one’s mouth when one is yawning. This norm does not constitute a new type of behavior, it merely regulates a behavior that we share, for instance, with chimpanzees. While it is unlikely that chimpanzees will want to cover their mouths too when they are yawning, there is nothing in Searle’s theory that would rule out this possibility. And so the question remains why chimpanzees could not also adopt this rule as a social norm.

There is also a second consideration that would seem to weaken Searle’s line of argument. Suppose one could show that institutional facts, as Searle describes them, can form a basis for social norms. Why could there not be something very similar in a complex society of animals? Many animals are capable of extended cooperative behavior. Why should this not allow them to assign a “value” to objects by following

[1]Thanks to our referees for pressing us on this point.

socially determined norm? The prohibition of infanticide that Andrews mentions seems to be just a case like that. The rule against infanticide ascribes a special value to the infants, and those who do not recognize this value are punished. Perhaps this is not a "status function" in the technical sense in which Searle uses the term. It is nevertheless a function that has no physical, but a social foundation. Why could such functional ascriptions not serve as a basis for social norms?

We are therefore willing to admit that Searle's theory of status functions does not provide a complete argument for claiming that social norms are a distinctively human phenomenon because they require an assignment of *status* functions in contrast to an assignment of functions based on their physical properties. Yet, we believe that Searle's theory of institutional facts provides a good starting point for developing such an argument. Just think of how easy it is for Searle to explain why only humans have invented money as a social good. It is Searle's prime example of a functional ascription that requires continued cooperation. When humans invented money, they did so by cooperatively creating a status function of the form "*X* counts as *Y* in context *C*". In this way, a certain physical object *X* (for example a piece of paper) acquires a new status (for example the status of money) to which a function is attached by way of collective intentionality. Nothing like this, Searle claims, can be found in the animal kingdom. For animals, no object can acquire a status that goes beyond the brute physical functions that they can assign to physical objects. (Searle, 1995, p. 40)

We will come back to Searle's view at the end of this paper. Let us now consider a different line of argument against the view that some non-human animals exhibit a primitive form of social normative behavior. Tomasello rejects this view on the ground that it confuses principles of instrumental rationality with principles of genuine normativity:

> Great apes may experience "instrumental pressure", for example, when they have a goal to eat food and they know that food is available at location *X*; this implies that they "must" go to location *X*. But this is just the way control systems with individual intentionality work: a mismatch between goal and perceived reality motivates action. In contrast, early humans began to self-monitor from the perspective of others and, indeed, self-regulated their behavioral decisions with others' evaluations in mind. (Tomasello, 2014, 74–75)

Let us bracket for a moment the question what Tomasello means by "self-monitoring from the perspective of others". The claim is that chimpanzees, when they show an apparently moral behavior like in Andrews' example, only experience "instrumental pressure". They follow principles of *individual* practical rationality like "If I want to get food, I 'must' go to location *X*" or "If I don't want to get in trouble with the mother of this infant, I 'should' be tolerant but not aggressive against it". What chimps are not able to understand, according to Tomasello's claim, are social principles like "I 'should' behave in this but not in the other way, because this is *what others expect from me*".

The question now is: How can we know that animals do not act in order to satisfy the expectations of others? This is the crucial step in Tomasello's argument. We would therefore like to quote him at length on this point:

> Finally, with respect to self-monitoring, the key is that being able to operate in this way communicatively requires individuals to self-monitor in a new way. As opposed to apes' cognitive self-monitoring, this new way was social. Specifically, as an individual was communicating with another, he was simultaneously imagining himself in the role of the recipient attempting to comprehend him (Mead, 1934). And so was born a new kind of self-monitoring in which communicators simulated the perspective of the recipient as a kind of check on whether the communicative act was well formulated and so was likely to be understood. This is not totally unlike the concern for self-image characteristic of early humans in which individuals simulate how they are being judged by others for their cooperativeness – it is just that in this case what is being evaluated is comprehensibility. Importantly, both of these kinds of self-monitoring are "normative" in a second-personal way: the agent is evaluating his or her own behavior from the perspective of how other social agents will evaluate it. [...] This social self-monitoring for intelligibility in cooperative communication lays the foundation for modern human norms of social rationality. (Tomasello, 2014, 58)

Can we conclude from this reasoning that animals do not understand social norms? It seems that Tomasello's argument warrants this conclusion if one accepts that a sharp distinction can be made between *individual* and *social* principles of rationality. But there are reasons speaking against such a clear dividing line, and this opens a way for defending Andrews' position. Take again the case of potato-washing mentioned in the quote from Searle earlier. If a chimp follows the principle "If I want to eat this dirty potato, I should wash it first", does he thereby follow a principle of individual or of social rationality? Formally, it is a principle of what he should do that has no implications about what others should do. But it is certainly important that this rule has been socially learned. And by "socially learned", we do not mean that one animal hands on knowledge to another one. It is a form of public learning. Whoever watches may pick up this rule and begin to follow it. So, it could be that a higher ranked animal in the group observes who is picking up the rule and who does not. This observation could have consequences for the social status of those who adopt the practice. They may be considered to be smarter and better partners.

Tomasello might still argue that this is not sufficient for reaching the level of "self-monitoring from the point of others". Yet, it shows that there are quite sophisticated forms of social intelligence that do not require such a form of socially induced self-monitoring. And so the question remains, whether these forms of social behavior might provide the basis for a primitive form of normative understanding.

10.4 The Case of Children

So far, we have argued that the conception of a primitive sense of normativity that Andrews advocates and that includes an implicit understanding of social norms may be defensible. An advocate of this conception can counter the objections of Searle and Tomasello. We now want to argue, however, that there are still other problems with this conception. In order to explain what these remaining problems are, we turn now to recent studies on the early understanding of normativity in human children. These studies have been partly inspired by Searle's and Tomasello's work, but they take us a significant step further, as we shall try to show now.[2]

A growing body of developmental studies suggests that even toddlers possess a normative awareness that manifests itself in different settings, notably in conventional and in pretend games, with various competences, for example, dealing with property rights, with artifact functions, or with various kinds of entitlements. The main evidence in these studies is that children show signs of protest, presumably in reaction to the violation of a rule.

In a seminal paper, Rakoczy, Warneken, and Tomasello (2008) proposed a new way of testing children's early understanding of norms. A basic idea underlying their approach is that norm violations may be perceived as a reason for protest, irrespective of what kind of norm is at stake. Hence, protesting could be a reliable indicator of normative understanding also outside a moral context.

In order to measure this basis awareness of normativity, Rakoczy et al. observe children's reactions to a puppet's violation of a constitutive rule in a conventional game. The children tested are between 2 and 3 years old. They are first familiarized with two novel actions, for example:

- In the model phase an adult shows 2- and 3-year-old children new game actions (*X* counts as *Y* in *C*). The adult performs actions *A*1 and *A*2. *A*1 is marked as "daxing", *A*2 as an accidental mistake
- In the action phase it is the child's turn to play the game of daxing, and to learn how to dax
- In the test phase, a third person (a puppet) enters and announces: "I'm gonna dax now!"
- In the target condition, the puppet performs an action which is mistaken, given the structure of the game
- Children's responses to such mistaken actions, in particular protest and correction, are taken as indicators of their awareness of the rule structure of the game

According to Rakoczy et al., the 3-year-olds saw the puppet's actions as not conforming to the social norm of daxing, and enforced the norm. Hence, the experiments are taken to show that 3-year-olds understand social norms. "These studies demonstrate in a particularly strong way that even very young children have some grasp of the normative structure of conventional activities" (Rakoczy et al., 2008). This conclusion is based on the background assumption that social norms have a foundation in the assignment of status functions. Therefore it is crucial that children are informed about a new game in which familiar objects have to be used in accordance with arbitrary rules. In this way, Rakoczy et al. think these objects acquire a status function and that it is therefore a normative requirement that one should use these objects only in the way prescribed by the rules of "daxing".

[2]The following sections are based on previously published work in cooperation with Beate Priewasser and Eva Rafetseder (see Brandl, Esken, Priewasser, & Rafetseder, 2015).

Could one explain children's protest also as a response to social pressure? Do children experience in these experiments such a pressure or only the pressure of individual rationality? Let us quote Tomasello once more in order to illustrate the question we are asking here:

> Young human children are concerned with the social evaluation of others from preschool years on, as they attempt to actively manage the impression they are making on them. [. . .]From the point of view of normativity, this meant that in making their behavioral decisions, humans not only experienced individual instrumental pressure but also experienced second-personal social pressure from their partners in social engagements. This constitutes one origin of what later become social norms of morality. (Tomasello, 2014, 75)

In the daxing-case, the child itself perfectly knows what it is supposed to do when it is her turn to "dax". It is therefore not the child that is under any kind of "pressure". Rather, it is the puppet that violates the daxing-rule that is put under pressure by the child. The role of the child is not to *experience*, but to let others experience a discomfort that is meant to change their behavior. We therefore need to put our question this way: Do children in these studies use their power as social agents to exert a "second personal pressure" on the puppet when it breaks the rules, or do they use their pressure merely to express their concern about a violation of some principle of individual rationality? Or do we have here another borderline case?

While we do not want to deny that the reported reactions may be indicative of a general awareness that something has gone "wrong", or that something "wrong" has been done, it would be premature to call these responses a "normative protest" merely for this reason. As soon as there is any evidence that a young child, an animal or an artificial system learns to distinguish between "right" and "wrong", we may attribute to it a basic awareness of normativity. One could define normativity simply as a standard that allows us to distinguish between "right" and "wrong", correct/incorrect, pleasant/unpleasant, expected/unexpected, or whatever. However, such categorizations can be understood like any other classification as purely factual. Simply calling something "right" or "wrong" does not yet make it right or wrong in a normative sense. Therefore, children might merely follow a certain pattern or regularity, when they first learn to use such terms, without grasping the deontology that we associate with them.

Consider a baby that reliably shows signs of protest in various circumstances, e.g. when her bottle is too hot, when the lid of the bottle is blocked, when the bottle does not taste sweet enough, etc. In this way even a baby can distinguish between "right" and "wrong" and would therefore manifest some normative understanding if we use the term "normative" without any restrictions. But babies do not show the right kind of protest when we present them with rule-violations in conventional games like the "daxing"-game.

What we conclude from this example is that there are two very different kinds of protest:

(a) protesting against some condition that is perceived as wrong, as unpleasent, or as unexpected, etc.
(b) protesting against rule-violations in conventional games.

The baby mentioned above is capable of expressing protest only in sense (a), while the protest of children in the "daxing"-game is of the quite different kind (b). While we agree with Rakoczy et al. on this point, it still needs to be shown whether this suffices for demonstrating a form of *normative* protest.

10.5 Three Steps Towards Normative Understanding

The protest behavior of young children raises a similar problem of indeterminacy that we already encountered in the discussion of Andrews' notion of primitive normativity. This leads us back to the fundamental question what it means to understand social norms. In the case of Andrews' examples of primitive normativity, we have seen that the behavior of chimpanzees can be interpreted as following either individual or social principles of rationality. Likewise, the protesting studies with 2- and 3-year old children leave it open whether children at this age really protest against the violation of social norms or some principle of individual rationality. No appeal to an understanding of social norms is necessary, for example, when a child learns from his caregiver how to open a water-tap: "You have to do it in this way, not that way". If the protesting behavior can be explained in the same way, then their protest does not indicate anything about their understanding of social norms.

We now start a third and last attempt to clarify what a basic understanding of social norms could mean. Let us therefore go back once more to Tomasello's idea to ground such an understanding in what he calls the ability to "self-monitor from the perspective of others". Without any doubt, the 2- and 3-year old children in the "daxing"-games are acquainted with the experience of social pressure. They may already experience a basic form of shame when they engage in social self-monitoring. Why then is it that it is still unclear, as we tried to show, whether they protest against the violation of a social norm? Something needs to be added here to the explanation that Tomasello offers. We now want to suggest that for answering this question it is critical to consider the effect that the authority of the experimenter has on children's response.

The authority of the experimenter derives from the fact that he tells the child (and the puppet) how to play the game of daxing. So far, it is his authority that decides which moves in the game are correct (appropriate) or incorrect (inappropriate). While this requires a social interaction between the one who introduces the game and those who are supposed to follow its rules, it is not yet a form of interaction that requires an understanding of social norms. There is social pressure due to the instructions handed out by someone who has the authority to do so, and it therefore seems that Tomasello's condition for normative understanding is fulfilled. But that is not necessarily the case since social pressure does not exclusively arise when social norms are operative. Authorities are an independent source of social pressure, even when they do not act according to social norms.

In order to make sure that a child is sensitive to norms, we have to contrast the demands of an authority with what a social norm requests.

10.5.1 The Ronnie Situation

Consider a child, Ronnie, that cries when she is put to bed and the lights are turned off. Suppose that Ronnie is old enough to notice a watch at the wall and that she observes that it is not yet 8 o'clock. When her Mum appears, presumably to turn off the lights, Ronnie does not immediately protest. She turns to the watch – perhaps pointing at it – to pass on her observation that the time to go to sleep has not yet come. But when her Mum ignores this and turns off the light anyway, Ronnie starts to protest heavily.

This is a case, we believe, that helps us to see how normative understanding can manifest itself in protest behavior. The fact that Ronnie looks up to the watch when her Mum comes in, gives us good reason to say that she understands the normative force of the rule "When it is 8 o'clock, it is time to go to bed". It does not matter that Ronnie might also protest when her Mum turns off the lights in accordance with the rule. It is her observing the watch that indicates her knowledge about when the lights *should* be turned off.

At first, the Ronnie-example may seem to be a simple case in which merely regulative rules are operative. We want to argue, however, that even in this case there is room for applying Searle's famous distinction between regulative and constitutive rules (see Searle, 1969). As defined by Searle, regulative rules "regulate antecedently or independently existing forms of behavior" (1969, p. 33). In our example, the rule to brush one's teeth before going to bed would be an example of a regulative rule, because the brushing can be performed independently of the rule. But what about the rule "go to bed at 8"? Surely, one can go to bed at other times, and thus perform the action antecedently and independently of the given rule. And one may do other things at 8, instead of going to bed, just as one can go to bed without brushing one's teeth. If one takes the rule merely in this regulative sense, however, it would not lead to a clear manifestation of normative understanding. No such understanding is needed to behave according to the instructions of adults requesting that such rules are obeyed.

In our example, Ronnie seems to exhibit a normative understanding, however. He does this by insisting that he does not have to go to bed *now*, but only at 8, because that is what the rule says. The action he wants to perform is "going to bed at the right time", and *this* action does not exist independently of a rule that defines 8 o'clock as the right time to go to bed. For this reason, we think that the rule here counts as a constitutive rule, like the daxing rule or other conventional game rules. It is a constitutive rule because it generates a new social fact. It is not a naturally given fact that 8 o'clock is the right time to go to bed. (Without a rule that establishes this, 8 o'clock is neither the right, nor the wrong time to go to bed). If children know that these rules are not freely (idiosyncratically) invented by an adult, they recognize an authority-independent ("objective") fact that generates a norm: 8 o'clock counts as the time to go to bed in the context of the evening routines in Ronnie's family.

At this point we agree with Searle's claims regarding the irreducibility of deontic phenomena (Searle, 1995, p. 70). This does not mean to deny, however, that an understanding of normative facts builds on children's experience with social pressure from authorities. From very early on in their lives, children are confronted with authorities that tell them what to do or not to do: drink this, do not touch that, go to sleep, etc. They come to understand such commands and know which circumstances prompt authorities to express them. At some point, they will find out that there are constraints for these commands to *be justified*. Even authorities cannot issue commands simply at their own will. There are rules that *entitle* them to do so depending on the circumstances. If it is 8 o'clock, Mum is entitled to turn off the lights. Once children know that there are these constraints even for authorities, they learn to understand something about normative constraints.

Taking all this into account, we can now differentiate between three stages in social cognition leading up to an understanding of social norms. In presenting our model, we are aware that much more needs to be said in order to flesh out this model in detail. Also, it would be interesting to compare our model with similar models that have been proposed in the literature on moral development, including the classical theory of stages of moral development by Laurence Kohlberg (cf. Kohlberg, 1981, 1982). Although Kohlberg's model targets a cognitive development in much older children, beginning at the age of 9, we notice certain parallels between how Kohlberg describes the early stages of this development and our model. Thus, when Kohlberg claims that, at a pre-conventional stage, children initially operate with a sense of right and wrong that is based on what feels good or bad, this description fits with our observation that social creatures develop a sensitivity for norms even before they are able to grasp the normative and/or conventional character of social rules.[3]

10.5.2 A Three Stage Model of Normative Understanding

Stage 1: Social creatures adapt to regularities that have a social foundation.

At this stage we find, for example, chimpanzees that obey a rule against infanticide. For Andrews this is already a form of implicit understanding of social norms. However, the social foundation of this behavior only consists in the fact that chimpanzees learn from others and create the rules as a group. Following Tomasello's argument, this kind of behavior requires only individual and no social

[3]Among psychologists, it is not completely unusual to apply Kohlberg's ideas in devising a stage model of moral development, beginning already in infancy. See for instance the entry on "Moral Development" in the online Psychology Encyclopedia (de Mott & Daena, 2016).

pressure to adapt to an existing social order. The apes need not understand what others expect from them, they may be driven only by instrumental rationality.

Stage 2: Social creatures make things happen that *should* be the case, or prevent things from happening that *should not be* the case (according to what authorities demand).

Children reach this stage when they protest in the "daxing"-experiments. In contrast to stage 1, they are now aware that agents do not only pursue individual goals according to principles of instrumental rationality. They now become sensitive to social pressure and therefore begin to monitor their own behavior and the behavior of others in order to avoid such pressure. Rakoczy and Tomasello take this to imply a new form of *social practical rationality*. But for the reasons pointed out, their sensitivity to social pressure should not be confused with an understanding of social norms prescribing what one *should do*. (We say more about the distinction between what *should be* and what *should be done* below).

Stage 3: Social creatures recognize social rules and know that they apply in all situations of a certain kind (independently of what an authority dictates).

When children reach this third stage, they learn to understand rules *as* rules in an abstract sense. This means that they are able to grasp the rule as something that exists independently of a concrete situation and can therefore apply to a certain *kind* of situation. They are then able to act in such situations because *the rule* requires it, not because it would be unwise to act against an authority. That is, we submit, when children begin to understand the nature of social norms.

10.6 The Case of Robots

Let us now turn to the case of robots. How far can the above considerations help to sort out our conflicting intuitions about artificial agents?

Starting with stage 1, we want first to reach an agreement that robots are social agents at least in the sense that they can adapt to regularities with a social foundation. To illustrate this first point, consider the following example of an adaptive social behavior of robots: There is a vast body of work on algorithms and control methods for groups of decentralized cooperating robots, called a "swarm" or "collective". These algorithms are generally meant to control collectives of hundreds or even thousands of robots. Each robot has the basic capabilities required for a swarm robot. For example, hundreds of small robots are moving on a table and create together the letter *A*. They are equipped with an algorithm to control each other: if one of them stops at the wrong position on the table, the others are able to inform the misaligned robot about its mistake.[4]

[4]Cf. http://www.eecs.harvard.edu/ssr/projects/progSA/kilobot.html.

The controversial question is whether robots might also exhibit a form of social practical rationality that includes a sensitivity to social pressure (provided by authorities). One might argue that this would grant too much to an artificial agent on the ground that robots are not able to follow rules. However, there is a further distinction to be made here, since we must distinguish behavior at stage 2 from rule-following at stage 3. Following Wilfried Sellars, one might introduce at this point a distinction between "ought to dos" and "ought to bes":

> Pattern-behavior of such and such a kind ought to be exhibited by trainees, hence we, the trainers, ought to do this and that, as likely to bring it about that it is exhibited. (Sellars, 1974, 423).

The trainee here might be a dog, an infant, or a robot. As its trainers or educators, we set the rules how it ought to behave. But as long as the dog, the infant or the robot does not understand the normative force of our rule, we should not describe it as a stage 3 case of following a rule, abstracted from the demands of concrete authorities. According to Sellars, we operate with such an abstract notion of rule-following when we reason about what one ought *to do*. While the trainee at stage 2 may not be able to participate in such reasoning, he may nevertheless understand what ought *to be*, and in this sense he, she or it may be able to follow a given rule. Without making this distinction, we would also have to deny chimpanzees that they can follow social rules, and we take that to be an empirically established fact.

Doubts still remain whether robots actually can be social agents in the sense of stage 2. In this case they would have to engage in a form of social self-monitoring as described by Tomasello. That means they would have to consciously experience that others expect from them to behave in certain ways and they would have to be able to evaluate their own behavior from a second-person perspective. Even if there are forms of robot behavior that come close to such self-monitoring, it is still very much an open question whether one might not reduce their social behavior to stage 1. So far we do not know whether robots can have any conscious awareness of what should or should not be the case and what others expect from them.

An interesting case study here is provided by the robot "Leonardo", built by Cynthia Breazeal and Brian Scassellati from the MIT Media Lab (cf. https://en.wikipedia.org/wiki/Leonardo_(robot).[5] Some researchers claim that Leonardo has a rudimentary theory of mind in the sense that it can model the beliefs and intentions of an interlocutor, including the interlocutor's beliefs about itself. But is such a description really warranted? Leonardo's motors, sensors, and cameras allow it to mimic human expressions, which helps humans react to the robot in a familiar way. Some authors describe it as mimicking human facial expressions, thereby distinguish between itself and others, and even to take the perspective of others. If this should mean that Leonardo understands something about the mental perspective of others and therefore possesses "mind-reading"-abilities, we strongly doubt this claim for the following reason. There is no evidence here that could not also

[5]Thanks to Johanna Seibt for drawing our attention to this case. For more about Leonardo and other social robots, see http://robotic.media.mit.edu.

be explained by saying that Leonardo keeps track of the interrelations between the facial expressions and the actions of an interlocutor, and in this way detects mismatches between them. It reads *behavior* but not *minds*.

At this point, the case of child development might be brought up as a further argument. Children experience social pressure from early on at an affective level, long before they develop secondary emotions like shame or pride. At this stage, social self-monitoring manifests itself as an awareness of being the center of attention of others. A child may feel uncomfortable (or secure) at an affective level without understanding what it is that makes her feel that way. When we try to transfer this non-conceptual consciousness of social relationships to the case of robots, we face the problem how we should justify ascribing any form of consciousness to artificial systems. But even if we set aside this problem, there is reason to remain skeptical at this point. We also do not know what kind of evidence would be sufficient to find a connection between social self-monitoring and affective consciousness in non-human animals. Yet, if we granted robots the ability to experience social pressure, we would have to grant this also to dogs and to apes that have been enculturated for several generations. Given the difficulties in answering this question in the case of animals, it is unlikely that research on artificial systems will be able to provide this kind of evidence in the near future.

Finally, we see no way how a swarm of robots would ever meet the conditions necessary for reaching stage 3. Not only are members of such a swarm probably not experiencing any social pressure (stage 2), they are hardly in a position to understand rules *as* rules, i.e. as abstract principles (in contrast to observable processes and abstractions from them) that possess an independent ("objective") normative force. We doubt that robots will ever be able to fulfill this condition.

Our conclusion of this paper is therefore a skeptical one. There is no evidence in sight warranting the claim that robots are social agents in some strong sense that requires an understanding of social norms. Despite this negative conclusion, our investigation also contains a positive suggestion how to make sense of the social behavior of robots. In our view, we may conceive of it in the same way in which we conceive of our social interaction with non-human animals or with infants when we train or educate them. In this case, we set the rules and bring it about that other agents conform to them. We enable them to adapt their behavior to our rules even before they can understand social norms.

Acknowledgements We are grateful for very helpful and substantial comments by Johanna Seibt, Raul Hakli and two anonymous reviewers. Research for this paper and the cooperation of authors was financially supported by Austrian Science Fund Project I638-G15, "Rule-understanding, shared intentionality, and the evaluation by others," as part of the ESF EUROCORES Programme EuroUnderstanding initiative.

References

Andrews, K. (2009). Understanding norms without a theory of mind. *Inquiry, 52*(5), 433–448.

Andrews, K. (2013). Ape Autonomy? Social norms in other species. In K. Petrus & M. Wild (Eds.), *Animal minds & animal ethics* (pp. 173–196).

Brandl, J., Esken, F., Priewasser, B., & Rafetseder, E. (2015). Young children's protest: What it can (not) tell us about early normative understanding. *Phenomenology and the Cognitive Sciences, 14*(4), 719–740.

de Mott, D., & Daena, K. (2016). Moral development. Retrieved from http://psychology.jrank.org/pages/431/Moral-Development.html (Psychology Encyclopedia, retrieved 29 July 2016).

Ginsborg, H. (2011). Primitive normativity and skepticism about rules. *The Journal of Philosophy, CVIII*(5), 227–254.

Kohlberg, L. (1981). *Essays on moral development* (The philosophy of moral development, Vol. I). San Francisco, CA: Harper & Row.

Kohlberg, L. (1982). Moral development. In J. Broughton & D. Freeman-Moir (Eds.), *The cognitive developmental psychology of James Mark Baldwin: Current theory and research in genetic epistemology*. Freeman-Moir, NJ: Ablex Publishing Corp.

Mead, G. H. (1934). *Mind, self, and society*. Chicago: Chicago University Press.

Rakoczy, H., Warneken, F., & Tomasello, M. (2008). The sources of normativity: Young children's awareness of the normative structure of games. *Developmental Psychology, 44*(3), 875–881.

Searle, J. R. (1969). *Speech acts: An essay in the philosophy of language*. Cambridge: Cambridge University Press.

Searle, J. R. (1995). *The construction of social reality*. New York: The Free Press.

Searle, J. R. (2010). *Making the social world: The structure of human civilization*. Oxford: Oxford University Press.

Sellars, W. (1974). Meaning as functional classification. *Synthese, 27*, 417–437.

Tomasello, M. (2014). *A natural history of human thinking*. Cambridge, MA: Harvard University Press.

Von Rohr, C., Burkart, J., & van Schaik, C. (2011). Evolutionary precursors of social norms in chimpanzees: A new approach. *Biology and Philosophy, 26*(1), 1–30.

Chapter 11
"I Tech Care": How Healthcare Robotics Can Change the Future of Love, Solidarity, and Responsibility

Antonio Carnevale

Abstract This paper is a continuation of a previous work where I argued that social expectations driving researches and developments in care robotics are not based on the simplistic optimism in a new commercial artifact but, involve a meaningful transformation of two rational features of human reality – ontology and normativity. In this paper my aim is to take a step forwards by investigating the kind of care relationship that could exist between a robot and a human. Usually, we take care of things and people because we love them, or else we want to give them support in their suffering. However, continuing to only value this sense of care, in a future rendered increasingly transparent and abstract by technology, may mean losing sight of the fact that taking care of others also means taking care of ourselves. If we completely entrust robots with the role of caring, the bigger concern is not the foreseeable decrease in the "humanity" in healthcare contexts, but the much more challenging notion of people surrendering the value and meaning in their lives. Since caring about something means, firstly, giving it value, a society passively nursed by technology is a society unable to give value to things and people. In order to avoid this risk, new approaches are required, no longer based on love or solidarity, but responsibility. I have named this approach "I tech care" and this paper aims to provide a general overview of the main concepts involved.

Keywords Human care • Love • Solidarity • Responsibility • Critical theory • Robotic technologies

11.1 The Power of Abstraction of New Technologies: "Transparency" and the "Moralization of the Things"

The Enlightenment placed the knowing subject at the center of the world. Advances in technology have led to a world that is no longer made up of people, their actions and thoughts, but a technological and informational structure where every action and

A. Carnevale (✉)
Scuola Superiore Sant'Anna, Pisa, Italy
e-mail: a.carnevale@alumni.sssup.it

R. Hakli, J. Seibt (eds.), *Sociality and Normativity for Robots*, Studies in the Philosophy of Sociality 9, DOI 10.1007/978-3-319-53133-5_11

thought is translated into computational codes (Floridi, 2013). Human rationality[1] seems no longer determined by the exchange between the thinking subject and the world of thought. The thinker and the thing being thought are both determined by the technological apparatus that is used to produce knowledge. In other words, for there to be cognition between a subject and an object, the world within which they are placed has to be prepared, organized, and planned.

This preliminary organization of the world is the mission of modern technology. The technology of the classical world, and pre-industrial societies was founded in a discourse on the "causality" of things.[2] The rational meaning of artifacts was given by their instrumental ability to replicate or imitate nature. In the Renaissance, Michelangelo Buonarroti argued that works of art are not created in the marble, but removed from the marble. A work of art is already present in nature; all the artist has to do is to remove it with the chisel, in order to bring out its beauty and replicate the perfection of nature. In our technological world, the idea of imitating or reproducing nature no longer exists, because we have completely changed our understanding of nature, so that it has become technologized and an artifact. What is the nature of a smartwatch? To measure time or to forget time? A means to avoid wasting time or a pleasurable pastime?

For most of our day we think, feel, and act in a reality whose objectivity is formerly constructed to guide us in thinking, feeling, and doing the things we have to do. The reality produced by traditional technologies was much more stringent and functional. A coffee cup is made so that using it properly constitutes an immediate action: the hollow space where to put the coffee, the handle to take the cup when the coffee is hot. Instead, modern technology offers solutions that involve its user. Using a vending machine requires information to be processed into a practical (and

[1]I speak about "human rationality" because, in my view, the predicted dissemination of assistive and care robotics constitutes a hugely challenging topic not only for some rational cultures – as the western world and its epistemological paradigms – rather of human beings as such. Whether the Enlightenment was still a form of rationality developed in Europe and North America, thanks or because of globalization, today the worldwide techno-scientific cultures move toward a general convergence. According to a report commissioned by the U.S. National Science Foundation and Department of Commerce we are aiming for a "Converging Technologies for Improving Human Performance" (http://www.wtec.org/ConvergingTechnologies/Report/NBIC_report.pdf). I do not sustain that technological progress is in itself a factor of human development. On the contrary, in my view, without freedom and democracy technology alone does not improve the human condition (Sen, 1999). But I am also convinced that the socialization and globalization of technologies we are witnessing – and healthcare robots are an emblematic example – will expand regardless of cultures and religions. These two processes end up complementing and hybridizing old and new myths (Hughes, 2012).

[2]Aristotle, in his doctrine of causality, defined four different typologies of cause of a manufactured thing (*Metaphysics*, I, 3). Considering the classic example of the statue, the sculptor is the efficient cause, bronze the material cause, the form taken by the statue is the formal cause, and the worship of the deity is the final cause.

not immediate) explanation in order to achieve our purpose. Understanding the price of the drink, inserting the coin, interacting with the display to select the product, choosing the amount of sugar, etc.

Technology thus influences our perception and understanding of the world. It enables us to relate to what surrounds us, thus helping us to form our vision and our judgment about the "facts". Overall, therefore, new technologies are more to do with how the image of the world is produced than with the world itself. The technologies we use today say much more about us and how we experience things than do things in themselves.

However, it is evident that by interfering in the way we learn about the world, the value and the weight we give to technologies is not always the same and depends on the level of experience that we personally have with the technology in question. Don Ihde proposed four possible forms of relationship between human beings and technology (Ihde, 1990):

1. *"Embodiment relation"*: an example given by Ihde is eyeglasses or a telescope, which one looks through rather than being interested in the technology itself;
2. *"Hermeneutic relation"*: this involves the interpretation of the technology, an example is a thermometer; we have to interpret the data on the display before we can apply it to the world it refers to;
3. *"Alterity relation"*: technology is experienced as a being that is otherwise, or as Ihde describes, a "quasi-other". An example would be an intelligent robot.;
4. *"Background relation"*: unlike the first three relations which are technologies that require direct and focal attention, the final category is located at the periphery of human attention. For example, an automated home heating system does not require daily attention; however, it continues to shape the human experience by providing a warm environment.

If these are the levels of mediation, what is the real power that technology now has to influence our ways of living and thinking?

This section makes reference to two forms (passive) of power: "transparency" (Clark, 2003), and the "moralization of the things" (Verbeek, 2011). Then in Sect. 11.4, I will describe a third form of power (active) which is connected to the responsibilities of care.

The technologies that have the greatest impact on our lives are invisible and "transparent", we interact with them without paying much attention, limiting ourselves to a minimal or peripheral exchange. The term "transparent" refers to the difference that Andy Clark made between "transparent" and "opaque" technologies (Clark, 2003). Clark argues that human thought is embodied – both in a biological and technological sense. The mind tends by nature to move towards something, to discover new areas of opportunity. However, in a technological society, the biological and finite nature of our brain does not allow this need for discovery to be fulfilled. Our minds are primed to seek out and incorporate non-biological resources, so that we actually think and feel through our best technologies.

We thus need technologies that help us to "think and feel through", that is, to build artificial conditions so that the mind can create a reality where no animal and biological life form has previously ever existed. Transparent technology does not change the human being by intervening on the physicality of the body, but because it makes us, according to Clark, "natural cyborgs". A cyborg is not a monster that is half man and half machine. There is no need for cochlear implants or prostheses to be a cyborg. We are natural cyborgs because we live inside the transparent prostheses of ourselves, because every day we use technological tools and languages that change our way of seeing each other and how we situate ourselves.

But technology is not only transparent, it is also made up of real and tangible devices that we consciously use every day. Phones, cars, computers, robots, and the like are tangible things with which we interact, and that contribute to our values and moral sentiments (Coeckelbergh, 2010). For example, the feeling of happiness of having a child would be impossible in a sterile couple without reproductive technologies. As claimed by the philosopher Peter-Paul Verbeek, the fact that technological objects do not have intentionality and consciousness in the human sense, does not mean that they cannot be attributed to some kind of moral consideration: "The ways in which technologies help us to act morally do not have the character of determination; technologies help to organize a relation between human beings and reality, which is the basis for certain moral actions and decisions and which depends on characteristics of both the mediating technology and the forms of their appropriation by human beings" (Verbeek, 2011, 141).

What conclusions can we draw from these two forms of technological power?

The history of human thought has accustomed us to believe that the place of the abstract, of the conceptual, was the reasoning in our mind. The intellectual and mental capability for "clarity and distinction". An example is the "*cogito ergo sum*" in René Descartes. On the other hand, we have also always believed that the physical and empirical reality was outside, the external world with its concreteness and opacity. Today, things no longer seem to be the same. In a world rendered increasingly transparent and moralized by technology, there is much more abstraction outside of us than there is in our heads. Taking inspiration from Marxist language, it can be said that technologies in reality create the condition for a "real abstraction", which is not logical (i.e., still situated in the minds of those who think things), but an abstraction which is born from differences and the mediation between differences, because it is capable of articulating an entire plan of society and fully realizing it.[3] Abstracting an idea from the reality is not merely a logical operation ($A = A$; $A \neq B$, etc.), but also involves a choice, therefore freedom. Abstracting means distinguishing, which means choosing, which in itself means being free to choose. Without freedom of choice there is no abstraction, but only captivity and

[3]In this conception of "real abstraction" I was inspired by the work of Roberto Finelli and his reinterpretation of the link between Hegelian and Marxist philosophies (see Finelli, 1999).

reification.[4] Paradoxically, the more technologies take over the abstraction of the world, the more we as human beings will be called upon to rethink our role on this planet.

11.2 Robots and Healthcare

How do healthcare robots deal with the technological powers of "transparency" and the "moralization of the things"?

In WALL·E, an animated film released in 2008, humans have escaped from the earth, made unlivable because of pollution, and have been living for 700 years on a mega spaceship, Axiom. Here there are robots that help human beings with everything. Axiom is equipped with every possible comfort, where people survive without working, without fulfilling the minimum physical effort, to the point of becoming obese and unable to walk independently. On Axiom people move on floating chairs (never walking) through which they can also feed themselves and communicate through holographic screens. Pure fiction or the prediction of an inauspicious future?

That robots are ready to enter into the private lives of people is a fact that will soon be reality. In the last decade, healthcare robots have been the main focus of several projects and prototypes conceived to improve quality of life and independent living, thus promoting an active and healthy ageing and reducing health and social costs. Robotic service solutions range from the simplest tele-presence functionalities to support caregivers, such as the Giraff developed in the ExCITE project (Giraff Technologies company, website: www.giraff.org), AVA (iRobot/AVA company, website: www.irobot.com/ava) and Luna (Ackerman & Guizzo, 2011), to the most complex, such as assistance for daily living activities (DOMEO project, website: www.aal-domeo.eu), self-management of chronic diseases, well-being and integration in a smart environment (Simonov, Frisiello, & Bazzani, 2012). There are also robotic applications that deal with social services in other environments, such as the garbage collection performed by DustCart (Ferri, Manzi, Salvini, Mazzolai, Laschi, & Dario, 2011), assistance in shopping centres (Kanda, Shiomi, Miyashita, Ishiguro, & Hagita, 2009) and smart office buildings (Veloso et al., 2012).

It could be argued that so far service robots are merely prototypes designed and implemented by visionary researchers, thus machines that will never meet a real market. However this is not the case. In May 2012, corporate recruiters and staffing firms posted over 6,000 online job ads for robotics skills, according to Wanted

[4]It could be argued, at least, that without freedom no level of abstraction is possible, but I would define this kind of abstraction – recalling one of Hegel's conceptions – a sort of "bad infinity" (*schlechte Unendlichkeit*), that is simply the finite repeated infinitely, an abstraction that is always realized in half, a representation of the world that denies itself and is forced to resort to false images to explain the contents of the real (*Encyclopaedia*, §93).

Analytics. Hiring demand for this skill set has grown steadily over the years 2010–2012. The data of May 2012 represents a 29% year-over-year growth compared to May 2011 and twice the volume of job ads seen online in May 2010. In May 2012, the majority of job listings were for engineering and healthcare positions. More than 2,000 jobs were advertised online for healthcare practitioners, the most of any occupational field, at a 46% year-over-year increase. In the same years, engineering related jobs that required robotics skills also increased, up 13% from May 2011.[5]

It is therefore quite clear that when these prototypes are marketed, in a few years we might be faced with such customization and personalization that we will no longer be able to see them just as simple "machines". Clinical studies have already shown that, in technologically-advanced societies, the desire to have relationships with others – a desire that, in anthropological terms, make us social animals – is being systematically replaced by fetishes and non-human partners (Levy, 2007). The title of a recent work by Sherry Turkle is indicative of this: *Alone Together: Why We Expect More from Technology and Less from Each Other* (2011). In industrial societies the repressed desire of the other was synthetically satisfied through the use of chemicals, i.e. so-called "drugs" that enabled the artifice of a reality to be lived that was altered and separated from the rest of the world, from others. In contrast, in post-industrial societies, the repressed desire is no longer satisfied by running away from reality, but by staying frantically attached to it.

In a world where technology is no longer purely the domain of industry but is now colonizing our private lives, self-realization is the new idol along with the fetish to worship. The "altered" state of consciousness no longer consists in being "outside oneself", but in the frantic search for an opportunity, even virtual, to really count, to be someone for someone else. Paradise is neither above us nor within us, but *between* us. This also explains the insurgence of new pathological addictions – such as Internet addiction (Young, 1996).

Now let's imagine what will happen as soon as healthcare robots become widespread, i.e. when autonomous machines are capable of taking care of human's most vulnerable aspects, behind which the most fragile traits of our personalities are hidden. The temptation to rely completely on them would be great. Especially in western societies there is a tendency to judge those aspects of life that involve being cared for as unproductive and deficient: illness, dependence, aging (Kittay, 1999; Nussbaum, 2006). The robots would become the panacea of every evil! (Sharkey & Sharkey, 2011).

[5]Retrieved from: http://moodle.jesuitcp.org/pluginfile.php/119027/mod_resource/content/1/robotic%20hiring.pdf.

11.3 Why Is Taking Care So Important for Human Beings? Love and Solidarity

Why all this focus on the relationship between care and robotics? What makes the human practice of taking care of things and people so special? What does it really mean when we say we care about something or someone?

Let's start with the basics. We take care of things and people because we love them (Kittay, 1999; Frankfurt, 2004). Between care and love there is an indissoluble partnership. We have a strong interest that those we love stay well. We are interested in their wellbeing so that they can thus continue to exist in order to receive our feeling of love. Because their existence and presence is not a matter of indifference to us. For a lover, the object of love is in the world in order for the lover to be able to exchange love from this object of love. They are objects in so far they fulfill our need to love. This is why it is said that love is not chosen. We do not choose to love our children; we love them "regardless". Underlying that "regardless", there is an emotional mechanism that emphasizes the quality of the objects of our love in order to preserve our need to love. "Love is not love/Which alters when it alteration finds/Or bends with the remover to remove" (Shakespeare, sonnet 116).

Thus, the concern that we feel for those we love is above all a personal interest. Our status, our mental balance also depends on their wellbeing. This is why those who love often suffer more as result of the suffering of others than from their own suffering. There is therefore good reason to define this first type of caring for as "narcissistic". "Narcissistic" does not necessarily mean "selfish", as if there was a utilitarian calculation behind the actions of lovers. Indeed, those who take care of loved ones do it with great personal sacrifice; more often they donate their time and freedom for the good of those they love.

It should be specified that the act of sacrificing one's own life for the good of others is not a pure and simple gift (Mauss, 1967; Derrida, 1995). As previously mentioned, the person we love represents something really significant for us. He or she is the guarantor of our need to love. It is difficult, then, to think that those who sacrifice themselves for love do it purely for free. While, on the one hand, a utilitarian economy has no space for loving someone, on the other hand, however, at stake is an economy of feelings, i.e. an exchange dictated by the strong emotional relationship that binds us to the love of others. In loving others "regardless" actually hides the need to be recognized by others (Honneth, 1995). We love not to receive something in return, but for a higher reason. The gratuity of those who give themselves to others cannot be reciprocated with some particular object, but only by returning gratitude and equal love (Ricoeur, 2005). Love is not utilitarian not because of its total and absolute gratitude, but due to a different order of exchange. It does not ask for a counterpart since any exchangeable good would be valued as ephemeral: it would count for me in the same way as it would count for anyone else. Instead, love asks for recognition; love asks to be loved, a similar exclusivity. I am the only one who can deserve the gratitude and the love of the person I love and care about. Herein lies the anti-utilitarian and narcissistic reason for the loving interest in cherishing for things and people. When we take care in the name of love, the last thing we want is for it to be reciprocated. The idea of being reciprocated almost offends the person who does things for love. This is because essentially it would lead the lover to lose their position of superiority guaranteed by the gratitude that the other owes him.

However, the care does not end with its narcissistic manifestation. In the emotional mechanism that leads us to love someone, there is the same dynamic which, in other contexts and according to other premises, makes us take interest in much more abstract matters, for example the love for a nation, for human beings, jazz or a football team. In fact, they are all generic and abstract realities, without a body or face. The thoughtfulness in caring about abstract entities can be defined as "universal". I feel strongly about the fortunes of my nation even though I do not really know or love any of its members. However, I nevertheless feel bound to it. While at a narcissistic level, care and love are inseparable, in the universalistic interest, the bond with the object of my care needs justifying. The main reason why I feel connected to entities I do not really know, is *solidarity*.[6]

Solidarity, like love, implies a recognition. However, unlike love, it consists of a less direct acknowledgment. We are no longer in front of the gaze of the loved one, whose mere presence reminds us to provide assistance and care. In front of us there is an image, an idea, a common cause. While with love, the categories of "abstract", "general", "universal" represent the death of love itself, solidarity is nourished to the same extent as the bond with the other is all-embracing (Loretoni, 2015). Love is exclusive, solidarity is inclusive.

In the name of solidarity, people can be prompted to share causes with which apparently they have no real connection. Nobody forces me to be interested in them. I receive nothing in exchange for being interested in them. Maybe not even gratitude. And there is no legal institution that constrains me to provide solidarity to anyone. In a different way, this feeling of solidarity involves a moral obligation to act, which is commendable but not strictly binding. Solidarity is the reasonable recognition that there is common ground between human beings or beings in general, since I can also feel a moral obligation toward the destiny of non-human entities, such as animals, the earth, the universe, even poetry.

[6]The philosophical debate on the theme of solidarity is very broad. Therefore it is difficult to give a complete account. Generally speaking, the idea of solidarity I refer to owes very much to authors such as Rorty (1989), Honneth (1995), Habermas (1998), and Benhabib (2002). According to their views, the processes of socialization in contemporary society have little to do with conventional drives. Social cohesion is no longer a matter of membership to a family, a clan, a class. As already Durkheim predicted by analyzing the division of labor, the modern concept of solidarity requires no empathy or identity with others with whom we are in solidarity. Instead, it requires the acceptance of the other as other, an acknowledgment based on a discourse of justice and not on belonging or benevolence (Habermas, 1998). Solidarity, justice and care, then, become two sides of the same coin. By treating the other according to the norms of friendship, love and care, our humanity is declared, but also our individuality as a person. The moral categories that follow such interactions are responsibility, relationships, and participation. The corresponding moral feelings are love, care, sympathy and solidarity (Benhabib, 2002).

To sum up: the practice of care has both a narcissistic matrix and a universalist one. The movement is the same: *investing a thing, a person, an idea with a value.* The invested value tells us that, in the name of that thing, person, or idea, it is worthwhile to act, struggle, and sacrifice a part of oneself. Whether for the health of a family member or for defending the freedom of a population, what drives us to help, support, aid, love, fraternize, is what they *represent* for us.

The practice of care is perhaps the aspect of human life that makes us truly "human beings". This is why it is so important. We could continue to exist as numbers in monetary economics that maximize profits and relativize losses. And this is why there is no demise in sight for capitalism. We could continue to exist in societies in which machines are built to replace us to do difficult work, as in the spaceship Axiom. But what makes us truly human is taking care of things and people, which means giving value to reality.

In the practice of caring we rise above the selfishness of the economic exchange, over the camaraderie of small communities. We are something more than mere living beings. We are human beings because we give representation – i.e. value – to our lives. Without love and solidarity, life certainly would continue to exist, but it would have no value. It would not be chosen by people, but only passively experienced – a bare life (Agamben, 1998). And without being chosen, life would not even be free, because choosing means being free to choose.

11.4 The Challenge: The Responsibility to Provide Healthcare in a Transparent Technologized World

For decades it has been believed that the most advanced robotics design in the field of assistance and sociality was trying to replace all (the humanoid) or some parts (the cyborg) of the human body (Henry, 2014). In the future this centrality of the anthropological element is probably doomed. In fact, the gradual incursion of robotics with other technological and scientific sectors – ICT, AI, synthetic biology, digital fabrication – will almost certainly lead to new trends.

On the one hand, there is the interest of researchers and developers in imitating and reproducing not only the human body on its own, but biology and nature in a broader sense – which is represented here by the project of the robotic octopus (www.octopus-project.eu).

On the other hand, the anthropological element will not disappear completely, but will be greatly transformed. No longer will the body be like a biological machine created to mimic and replicate, but humanity construed as a unique normative element, i.e. a model driven by both biological and social rules. The degree of "humanity" of a machine will no longer be represented by its aesthetic and functional similarity with the human body, but by its ability to choose based on principles and shared rules (Wallach & Allen, 2009). The more complex machines become, in order to be more "human" they must also be "right", making decisions according to universalizable rules – like the well-known laws of Asimov. There are interesting legal approaches that imagine the rules to which technology should be subjected, not as rules to regulate the technical functionality of its product, rather as tools of rights. By regulating the use of technological artifacts, it is possible to intervene and improve some aspects of people's behavior (van den Berg, 2011; van den Berg & Leenes, 2013).

If this is the robotics of tomorrow, it is difficult to believe that in the future, the problem with healthcare robots will be their similarity to a pleasant and attentive caregiver. The robot does not necessarily have to love the person they are caring for, nor have solidarity for the cause of his/her suffering, nor look like a good mother or a loving pet. If it really will be possible to reproduce the feeling of "love" in the machine, this will still be a matter of programming the commands and the rules that the robots have to follow, and not an ontological question about their sensitivity.

The problem is normative and techno-regulatory, and not speculative. Accordingly, the ensuing questions are thus: how will humans and machines be able to live together? In a world rendered transparent and moralized by technological processes, what will become of the significance of the human touch, an affectionate gaze, a hug? What is to become of the typical human feelings such as sympathy, guilt, shame, and even a sense of justice? If it is true that those who suffer injustice are more able to enjoy the taste of freedom, who will still have a sense of freedom in a society in which autonomous machines will do anything and everything to prevent us from suffering?

These questions highlight the real challenge that the future diffusion of healthcare robots poses: the responsibility of providing a healthcare balance between technology and human values. In Sect. 11.1 we saw two negative forms of the power of technology: firstly, the power of transparency which invisibly affects people's behavior, and secondly the power that technological artifacts have to influence our moral behavior through our interaction with them. This is the reason they are negative powers: they decentralize the intentionality and therefore the responsibility of the people and users. Both when the technology works invisibly, and when it is tangible, human agency is to some extent determined by technological agency. But our relationship with technology is not only negative. Healthcare Robotics is an example that helps us understand that technology, in addition to being transparent, it has an important additional power which the individual is fully aware of i.e. the power to choose between health and disease.

A few years ago Angelina Jolie, the famous actress, had a mastectomy for fear of contracting breast cancer, and more recently decided to have her ovaries removed for the same reason. Her decisions were made in the light of a tragic family history, with the deaths of her mother, grandmother and aunt to cancer. This example illustrates how scientific research and genetic testing techniques on the probability of countering cancer force us to make a choice. The more technological development involves specific aspects of our biological and psychological life, the more it produces an order of reflections that puts us between "disease" and "health" – and therefore an order of reflection that makes any concept of human nature relative (Buchanan, 2009).

However, this power of technology to position ourselves somewhere between sickness and health can be read in two ways. From a negative point of view, one might think of some kind of constant exposure to a disease. Because we are all a little sick, we have to depend on health institutions and the bio-political power of medicine. However, there is also a positive reading: i.e. the approach that I call *I tech care*. This approach does not depend on a doctor taking our place and telling to us what is good or bad for us. By placing us between illness and health, technology creates the condition of "not-yet-ill" (Verbeek, 2011). This means that we have the opportunity to be able to know in advance the likelihood of suffering, of getting sick, and this gives us to the possibility to intervene in our health status, even by choosing to have one of our organs removed. This ability gives us a certain degree of responsibility regarding our way of life.

In other words, in an *I tech care* approach, a responsible health condition is not a condition that does not ignore the fact that one might contract a disease, nor is it a condition that fatalistically accepts the possibility of getting sick, but rather a condition that we contribute to choosing, and which puts us in a continuous relationship between sickness and health. It is a difficult philosophical and political challenge. Scientific progress cannot and should not be stopped. However it is unacceptable that human freedom should have to adapt to the needs of technology. In this paper I do not define the details of the approach that I am proposing, for the moment I limit myself to outlining its major characteristics.

- *Contextual and a case-by-case understanding.* Firstly, approaches are needed that place the subject of care at the center. The person needing to be cared for is the person who should be asked what they need to strengthen their personal autonomy. However, to understand the perspective of those seeking assistance, their context also needs investigating. The links between those who offer care and those who receive it are so intriguing and binding that they need explaining. In order to offer a better public service, healthcare agencies should understand whether behind the demand for care there is a relationship motivated exclusively by love or (qualified) work, or whether there is a context of asymmetry, exploitation, or circumvention.
- *Thoughtful.* Secondly, approaches are needed that help to create a spirit of reflection. In other words, a future technologized society needs to promote the awareness that dealing with care is not a matter of compassion and benevolence, but a wide-ranging responsibility. Caring about others also means caring about ourselves. The topic of care needs to move to a higher stage of public attention through public debates in the media, through education in secondary and high schools, higher education and training in universities, in interdisciplinary field of research in postgraduate training. This would help create a common technological culture that does see the benefits of care technologies only in financial terms, but places a strong emphasis on the interaction between technology and values, reflecting on the importance of living and cooperating together to try to solve problems and dilemmas (Pirni & Carnevale, 2014).

- *Critical.* Finally, approaches are needed that foster a critical attitude in people and institutions. By "critical", I do not necessarily mean an attitude that is a priori contrary to the development of science and technology. Being a priori against – as well as being totally in favour of – techno-scientific development means thinking in mono blocks, as if technology and science were animated by an evolutionary rationality able to pursue its own purposes. But this is not the case. As previously mentioned at the beginning of this article, technology will never again have to do exclusively with the purpose of things. Rather it will be a *"language of the world"* that can transfer the abstract concepts of society into reality. Technological means and human purposes will therefore increasingly be interchangeable concepts. As a result, a "critical" attitude means being able to match the best means with the best purposes.

Encouraging this attitude also has a second benefit. Critical could involve the capacity *also* to choose traditional caring practices when they are deemed effective and better performing than technologically advanced solutions. In summary, a critical approach knows how and when to choose the future, but knows also to go back, stopping when it is inappropriate. There is no need to prefer the new at all costs. The future should be a change not a necessity. To makes this change requires stronger moral and ethical arguments, which are more comprehensive as well as convincing in order to help us to make distinctions and decisions to find the right balance between means and ends, case by case (Battaglia & Carnevale, 2014). Although human values and technology always need to be guided by pragmatic, contextual, and never dogmatic ethics (Lucivero, Swierstra, & Boenink, 2011), it is also true that we can share experience only on the basis of universal principles (Datteri & Tamburrini, 2009; Feil-Seifer & Matarić, 2011).

In conclusion, with the formula *I tech care*, my aim is to stress the importance of proactively accepting the challenge of healthcare technologies, which involves the way in which humans (a) attribute values to things and people, and (b) make their own abstraction of the world. This is no trivial challenge of responsibility and is almost unprecedented in modern times. We have the opportunity to rethink from scratch the moral and ethical obligations that modern institutions have created in favor of the most vulnerable or disadvantaged subjects. Namely, the institutions of the classic and industrial societies have mostly taken care of people to whom they were naturally or conventionally related. The family, clan, community, or social class. But in the post-industrial society, the ethical bonds have been put under pressure and, because of the capitalist system, today they tend to dissolve or become selfish and solipsistic. The challenge of assistive and care robotics is this: either our institutions are able to increase care, reasoning from the particular to the universal, from the loved person and the member of community to the human being as such, or we are destined to a society in which we no longer know how to take care of anything. This would be a society that is not free, because we consign the task of giving value to things and people to technologies.

We are actually up against a double challenge. On one side are the subjects, on the other the institutions.

The subjects demanding care are the first to be interested in the transformation of healthcare systems. Thus, before investing public money in robotic and technological solutions to the issue of care, it would be good to ask these subjects. Most likely we would discover that many features of the ideal decision-making model actually exist in practice. As "disability studies" have taught, being a subject demanding care does not mean being handicapped (Kristiansen, Vehmas, & Shakespeare, 2009). There is no correlation between the limitation of a physical or cognitive ability and being considered as deficient for the whole of humanity.

Vulnerable and aged people, if helped, are perfectly able to use technology responsibly to enhance their autonomy. Just to make an example, in 2011 A.T. Kearney's Global Maturing Consumer study, the first ever of its kind, interviewed 3,000 people older than 60 in 23 countries – from China to Britain, from Mexico to India, from the United States to Russia. The study shows that mature consumers have time and want to be well informed, so they tend to be heavy Internet users and shoppers. "Exactly half of the survey respondents use the Internet, with 20% taking advantage of it for shopping, research and communication." "The younger respondents, and the wealthier and older participants, say they tend to purchase more products online, and are also more likely to use home delivery services, particularly in Japan and the BRIC countries (Brazil, Russia, India and China). Mature Mexicans and Brazilians are particularly keen to shop online, and in the United Kingdom 'silver surfers' (55 and older) are the fastest-growing age group for Internet adoption."[7] This means that mature consumers are not afraid of technology.

Accordingly, investigations into the future implications of healthcare robots require some changes to the current notions of personal autonomy, which have been excessively inspired by liberal individualism (Kittay, 1999; Benhabib, 2002). Personal autonomy should be reconsidered as a capacity that exists only in the context of the social relations that support it and only in conjunction with the internal sense of autonomy: one's autonomy is vulnerable to being disregarded in one's relationship with others (Honneth, 1995).

However, in a world rendered transparent and moralized by technology, the responsibility to provide healthcare requires a consideration well beyond the role of the subjects. Mediating between technology and the duty to care, choosing the best technological means in view of the best human purposes, which are all issues of responsibility that lead to an institutional level (van Wynsberghe, 2013). The burden of caring cannot be entrusted just to voluntary associations and workers. In voluntary work, cherishing has a lot to do with personal responsibility: I give you my word that I will take care of your needs. I could also take care of people that I do not know, but the responsibility that guides me is always personal, a moral maxim: one should treat others as one would like to be treated by others. If this line of thought is not totally new in the history of moral and political philosophy, the historical situation in which we could soon find ourselves instead is absolutely new.

[7] Retrieved from: https://www.atkearney.com/it/health/ideas-insights/article/-/asset_publisher/LCcgOeS4t85g/content/what-do-mature-consumers-want-/10192.

With the increasing technologization of society, people and institutions are realizing something that philosophers and artists have often claimed: the sense of finitude and imperfection of the human condition. If one believes that robots will soon take care of our fragility, the first feeling is that of safety, protection. But it is only a sensation. Actually, the more society becomes technologized and human relationships become virtual and abstracted, the more this historical transition reveals the general character of human dependence.

This unexpected awareness constitutes the real challenge of healthcare robots: understanding that dependence is not a private question, but rather the universal representation of humanity (Carnevale, 2015). We are frail creatures from birth. We thus build societies and create technologies to overcome these anthropological difficulties. However in so doing we become universally vulnerable because social and political life requires us to be increasingly supported by artificial and conventional measures. In the future, independence and dependence will be similar concepts that are hard to separate.

11.5 Conclusions

Once it was philosophy that gave interpretation and structuring morally the world. It is likely that in the future we will expect technology to fulfill this role; the power of transparency and the moralization of the things that technology has to offer, force us to find a new position and a new sense of responsibility. In this historic transition – the thesis of this paper – healthcare robots constitute an emblematic example. The prospect of their future dissemination offers us the opportunity to totally rethink the moral and ethical obligations that modern institutions have created in reference to those who require care.

In a certain sense we have no choice in the matter. Either we will be able to increase the level and extent of care by reasoning from the particular to the universal, or we will be destined to live in a society in which we no longer know how to take care of anything. And this will thus be a society that is not free – because we will consign the task of giving value to things and people to technologies.

Acknowledgements This paper is a continuation of a previous work (Carnevale, 2014) and it is one of the outcomes of the Research Unit "Social Asymmetries and Political Inclusion: Concepts, methods, and policies", Scuola Superiore Sant'Anna, Pisa (Prin 2010–2011: Local Coordinator, Barbara Henry; National Coordinator, Laura Bazzicalupo).

References

Ackerman, E., & Guizzo, E. (2011). Mystery robot revealed: Robodynamics luna is fully programmable adult-size personal robot. *IEEE Spectrum*, 11 May 2011.

Agamben, G. (1998). *Homo sacer: Sovereign power and bare life* (D. Heller-Roazen, Trans.). Stanford: Stanford University Press.

Battaglia, F., & Carnevale, A. (Eds.). (2014). Reframing the debate on human enhancement. *Humana Mente Journal of Philosophical Studies*, issue 26.

Benhabib, S. (2002). *The claims of culture: Equality and diversity in the global era*. Princeton: Princeton University Press.

Buchanan, A. (2009). Human nature and enhancement. *Bioethics*, *23*(3), 141–150.

Carnevale, A. (2014). Ontology and normativity in the care-robot relationship. In J. Seibt, R. Hakli, & M. Nørskov (Eds.), *Sociable robots and the future of social relations: Proceedings of robo-philosophy 2014* (pp. 143–152). Amsterdam: IOS Press.

Carnevale, A. (2015). Robots, disability, and good human life. *Disability Studies Quarterly*, *35*(1). doi:http://dx.doi.org/10.18061/dsq.v35i1.4604

Clark, A. (2003). *Natural-born cyborg*. Oxford: Oxford University Press.

Coeckelbergh, M. (2010). Robot rights? Towards a social-relational justification of moral consideration. *Ethics and Information Technology*, *12*(3), 209–221.

Datteri, E., & Tamburrini, G. (2009). Ethical reflections on health care robots. In R. Capurro & M. Nagenborg (Eds.), *Ethics and robotics* (pp. 35–48). Amsterdam: IOS Press.

Derrida, J. (1995). *The gift of death* (D. Wills, Trans.) Chicago: University of Chicago Press.

Feil-Seifer, D. J., & Matarić, M. J. (2011). Ethical principles for socially assistive robotics. *IEEE Robotics & Automation Magazine*, *18*(1), 24–31.

Ferri, G., Manzi, A., Salvini, P., Mazzolai, B., Laschi, C., & Dario, P. (2011). Dust-Cart, an autonomous robot for door-to-door garbage collection: From DustBot project to the experimentation in the small town of Peccioli. In *2011 IEEE international conference on robotics and automation (ICRA)*, Shanghai (pp. 655–660).

Finelli, R. (1999). Abstraktion. In H. Sandkühler (Ed.), *Enzyklopädie philosophie* (pp. 20–23). Hamburg: Felix Meiner Verlag.

Floridi, L. (2013). *The ethics of information*. Oxford: Oxford University.

Frankfurt, H. (2004). *The reasons of love*. Princeton: Princeton University Press.

Habermas, J. (1998). *Inclusion of the other: Studies in political theory* (C. Cronin & P. D. Greiff, Trans.). Cambridge MA: MIT Press.

Henry, B. (2014). Imaginaries of the global age. 'golem and others' in the post-human condition. *Politica & Societá Periodico di filosofia politica e studi sociali*, *2*, 221–246.

Honneth, A. (1995). *The struggle for recognition: The moral grammar of social conflicts* (J. L1Anderson, Trans.). Cambridge: Polity Press.

Hughes, J. (2012). The politics of transhumanism and the techno-millennial imagination, 1626–2030. *Zygon: Journal of Religion and Science*, *47*, 757–776.

Ihde, D. (1990). *Technology and the lifeworld: From garden to earth*. Bloomington: Indiana University Press.

Kanda, T., Shiomi, M., Miyashita, Z., Ishiguro, H., & Hagita, N. (2009). An affective guide robot in a shopping mall. In *Proceedings of the 4th ACM/IEEE international conference on human robot interaction (HCI)* (pp. 173–180).

Kittay, E. F. (1999). *Love's labor: Essays on women, equality, and dependency*. New York: Routledge.

Kristiansen, K., Vehmas, S., & Shakespeare, T. (2009). *Arguing about disability: Philosophical perspectives*. London: Routledge.

Levy, D. (2007). *Love and sex with robots: The evolution of human-robot relationships*. New York: HarperCollins.

Loretoni, A. (2015). *Ampliare lo sguardo*. Rome: Donzelli.

Lucivero, F., Swierstra, T., & Boenink, M. (2011). Assessing expectations: Towards a toolbox for an ethics of emerging technologies. *NanoEthics*, *5*(2), 129–141.

Mauss, M. (1967). *The gift: Forms and functions of exchange in archaic societies* (I. Cunnison, Trans.). New York: Norton.

Nussbaum, M. C. (2006). *Frontiers of justice: Disability nationality species membership*. Cambridge MA: Harvard University Press.

Pirni, A., & Carnevale, A. (2014). Technologies change – do we change as well? on the link between technologies, self, and society. *Politica & Societá*, *2*, 173–184.

Ricoeur, P. (2005). *The course of recognition* (D. Pellauer Trans.). Cambridge MA: Harvard University Press.

Rorty, R. (1989). *Contingency irony and solidarity*. Cambridge, MA: Cambridge University Press.

Sen, A. (1999). *Development as freedom*. Oxford: Oxford University Press.

Sharkey, N., & Sharkey, A. (2011). The rights and wrongs of robot care. In P. Lin, K. Abney, & G. A. Bekey (Eds.), *Robot ethics: The ethical and social implications of robotics* (pp. 267–282). Cambridge MA: MIT Press.

Simonov, M., Frisiello, A., & Bazzani, M. (2012). Using humanoid robot in ambient-assisted living. *Global Telemedicine and eHealth Updates: Knowledge Resources*, *5*, 438–442.

Turkle, S. (2011). *Alone together: Why we expect more from technology and less from each other*. New York: Basic Books.

van den Berg, B. (2011). Robots as tools for techno-regulation. *Law, Innovation and Technology*, *3*(2), 317–332.

van den Berg, B., & Leenes, R. (2013). Abort, retry fail: Scoping techno-regulation and other techno-effects. *Human Law and Computer Law: Comparative Perspectives*, *25*, 67–87.

van Wynsberghe, A. L. (2013). Designing robots for care: Care centered value-sensitive design. *Science and Engineering Ethics*, *19*(2), 407–433.

Veloso, M., Biswas, J., Coltin, B., Rosenthal, S., Kollar T., Mericli, C.,... Ventura, R. (2012). Cobots: Collaborative robots servicing multi-floor buildings. In *2012 IEEE/RSJ international conference on intelligent robots and systems*, Vilamoura (pp. 5446–5447).

Verbeek, P. P. (2011). *Moralizing technology: Understanding and designing the morality of things*. Chicago: University of Chicago Press.

Wallach, W., & Allen, C. (2009). *Moral machines: Teaching robots right from wrong*. Oxford: Oxford University Press.

Young, K. (1996). Internet addiction: The emergence of a new clinical disorder *Cyberpsychology and Behavior*, *3*, 237–244.

Chapter 12
"Robot" as a Life-Form Word

Hans Bernhard Schmid

Abstract In much of earlier philosophy of robotics and artificial intelligence, it is argued that while robots can perform fully standardized routine work, they are, as a matter of principle, unable to participate in the discursive practices within which our social form of life is negotiated. With robots (and their virtual counterparts, the bots) currently entering our service economy, it is not entirely unlikely to assume that this view is about to be disproven by the facts. It may well be that robots will soon be more "like us" in that the conventions and practices that shape our social form of life will be negotiated, in part, in cooperation with and partly even among our mechanical helpers. This paper investigates into the reasons for and causes of some of the fears at this prospect – a field of investigation in which literary fiction has often been more clear-sighted than received philosophy of robotics. It is argued that an influential and historically deeply rooted form of "robophobia" originates in an "us" versus "them" frame of mind that rests on a concept of radical self-alienation of mankind.

Keywords Robophilosophy • Robophobia • Discursive practice • Life-form • Self-alienation

12.1 Introduction

Whatever definition of "robot" may eventually come to be generally accepted (if any at all), it is hard to imagine that it will not include some version of the idea of an artificial, mechanical substitute for natural agents, and especially (but not only) for humans. If this is true, the concept of "robot" obviously entails a great deal of evaluative aspects. "Substituting humans" comes with hopes and fears and engages the values and disvalues on which these and similar attitudes are focused. "Robot" is thus a comparatively "thick" concept – it is a matter of evaluation as much as of description, and robophilosophy thus cannot ignore the evaluative dimension. Or, to

H.B. Schmid (✉)
Department of Philosophy, University of Vienna, Universitätsstraße 7, A-1010, Vienna, Austria
e-mail: Hans.Bernhard.Schmid@univie.ac.at

R. Hakli, J. Seibt (eds.), *Sociality and Normativity for Robots*, Studies in the Philosophy of Sociality 9, DOI 10.1007/978-3-319-53133-5_12

put it more colloquially: in order to find out something about our concept of "robot", we should also examine how we *feel* about robots.

Many of the feelings in question seem to be of the wary kind. There is some fascination with new technology, but genuine robo-enthusiasm is rare. It seems that we do not really *like* the idea of being replaced. If our current industrial robots are widely accepted or even acclaimed in the public discourse, it is perhaps not so much because we like the idea of robots taking assembly line work off our hands, but rather because this seems to be the only means of stopping the decades-long trend of de-industrialization in western countries in a globalized economy. The general sentiment is that it is better to accept the price of automatization than losing the secondary economic sector to developing countries altogether, where human labor is cheap. This seems to be the reason why the pictures of whole assembly lines that are practically devoid of human presence and where production is now largely in the "hands" of robots, do not seem to stir much public sentiment.

With recent developments of so-called "social" robot technology and artificial "social" intelligence, however, the public sentiment seems to be changing. We accept handing over the primary and secondary economic sector to robots, but as robots – with their virtual or software siblings, the bots – now seem ready to do service work and thus enter the tertiary sector as well, hence the domain of *interaction*, the debate takes on different tones. It seems to be o.k. for robots to serve us in *production*, but having them serve us in social *interaction* with them somehow raises a new kind of worry. Some degree of fear of robots now seems to be the dominant feeling – let us call it *robophobia*, and use the term in as neutral a way as possible, leaving open the question of whether robophobia expresses an exaggerated, inadequate, and ultimately irrational attitude, or an adequate sense of the danger implied in the potentials of robo-technology. As a recent and rather drastic example, Stephen Hawking is reported in global media to have issued the warning that "artificial intelligence could spell the end of the human race",[1] and with other prominent figures joining this most recent surge of robophobia, the press headlined: "Bill Gates says we *should* fear a robot uprising: Microsoft co-founder says he agrees with Elon Musk on dangers of AI."[2]

The wording of these and similar messages and the imagery used in these warnings are significant. Indeed it seems that the fear of a robot-uprising that ends the human race, hyperbolic as it may seem, is an idea that is built into the very conception of the "robot" right from the origin of the word in Karel Capek's 1920 play "Rossum's Universal Robots". Capek's play depicts a dystopian scenario in the future, in which artificial, human-like machines develop into a life-form that ultimately *displaces* mankind. As this imagery has so persistently stuck to the word "robot" in the public discourse, the kind of "robophilosophy" (a term coined by Johanna Seibt; cf. Seibt, Hakli, & Nørskov, 2014) that is sensitive to the evaluative

[1] http://www.bbc.com/news/technology-30290540

[2] http://www.dailymail.co.uk/sciencetech/article-2931375/Bill-Gates-says-fear-robot-uprising-Microsoft-founder-says-agrees-Elon-Musk-dangers-AI.html

dimensions of the concept is well-advised to examine it in some more detail. Science fiction and the sentiments it evokes may turn out to be another source for future robophilosophy, if perhaps not as important a source than hard facts about technology – because such stories seem to have shaped the concept.

Capek projects the time of action of his 1920 play a couple of decades into the future, presumably into the 1950s and 1960s. In this imagined future, robot technology has reached a level at which robots are cheap, in which economic activity is largely robotized, and where robots are able to behave in a very human-like fashion. Over the course of the time of the play (which spans over more than a decade), robot technology makes rapid further progress. At the beginning, they are – as stated in the stage directions – "slightly mechanical in their speech and movements", and "blank of expression, fixed in their gaze", but over the course of time, they come to display the full range of human capacities of interaction, thought, and emotion, culminating in a display of altruistic sentiment and indeed romantic love at the bizarre end of the play. Capek's core topic is the development of a robot life-form and the relation between *ours* and *theirs*. From being well-integrated in a serving role into our life-form as rather active participants, they come to develop *their own*, which ultimately (and rather violently) *displaces* ours.

The first act, in which robots are seen playing along in human economic activity swiftly and efficiently, negotiates the question of whether or not, given their human-like capacities and their economic role, they should be made full participants in the economy by being paid for their work. This view is advanced by an organization called the "League of Humanity", but it is rejected on two grounds: First, it is claimed that since robots are machines, they simply cannot be people, and that only people, and not machines, should be paid for work. Second and perhaps more importantly, it turns out that the robots do not *want* to be paid for their work, because they basically have no interest whatsoever in becoming consumers of the goods they produce, fully identifying with their role as *workers*. This ends the attempt to recognize robots as members of *our* community. The robots, however, come to see humans as parasites on them rather than seeing themselves in a symbiotic relation with the human consumers of their products. They rise in rebellion against humans and turn the table against them, finally killing all but one – their constructor – who is spared because even though he is a human, they recognize something very robot-like in him: "He is a worker", they say: 'He works with his hands like a robot. He builds houses. He can work." Moreover, they recognize that in order to maintain themselves, they will need him to enable them to reproduce. Capek's somber dystopia ends with an eerie ray of light in the epilogue, when the last human recognizes in two of the robots creatures with the capacity of compassion, self-sacrifice, and love, and hands things over to the Adam and Eve of a new robot life-form. Robophobia turns into – or culminates in – an uncanny sense of hope, and this mix of conflicting emotions concludes the play.

In Capek's, as in so many other cases in the history of sci-fi, the projected future has taken a somewhat different course. The 1950s and 1960s did not see robots of his kind. Instead, something very different and indeed contrary was developed, something that may somewhat alleviate the kind of robophobia initiated by Capek.

Influential skeptical arguments were put forth in philosophical research, arguments aimed at showing that creatures of the Capekian kind are in fact *impossible*, and that there thus cannot be any reason for robophobia of the Capekian kind. The very concept of a "universal robot" that can aspire to a life-form of its own was claimed to be an inconsistent idea, suggesting that any such dystopian scenario and the fears it involves only distract from the *real* issues of automatization that we should address. Michael Polanyi (1958) has initiated an influential line of critique, arguing that robot activity is limited to certain kinds of behavior. This view has been developed further to the claim – defended in a wide variety of versions in later literature – that however smart robots may become in the future, they cannot engage in the kinds of interactions that constitute a life-form of our kind. Robots, the claim goes, can do *standardized or routine work*, but they cannot engage in *discursive or normative practices* in which the form of life such as ours are negotiated and transmitted. As routine work is what it is only within a life-form, the kind of agency instantiated by robots is non-causally dependent on our life-form. From a different perspective, Paul Ziff (1959) has claimed that psychological concepts such as thought, feeling, and action can be applied to machines only equivocally because they are not – and by definition, cannot be – *living*. Whatever form the activity of future robots may come to have, it will not be a *life*-form.

The first line of defense emphasizes what robots *do*, the second line emphasizes what robots *are*, and both undermine the plausibility of the scenario within which the term "robot" originated and thus the plausibility that robo-technology may give a reason for robophobia of the Capekian kind. Neither of these skeptical views rules out in principle that mankind might somehow tragically end up being killed by some self-developed machinery, but in these views, no such event could ever be the result of a proper robot "uprising" or "revolt" – not to speak of human life being *replaced* by a new and independent robot life form along Capekian lines. This may seem to be cold comfort – if you're being killed by a robot, you may not care that much if you're being killed by a proper agent or just some simulation or an ontologically dependent version thereof. But the difference suggests an important difference in the *kind of danger* at stake, and thus does make a difference in the *kind of fear* that seems appropriate. Put in philosophical jargon, danger is the formal object of fear – it is in virtue of something's being seen as *dangerous* that the attitude towards that target is understandable as one of fear (rather than, say, one of hope). As fear comes in many forms, a difference in danger rationally requires a difference in fear. And it makes an obvious difference if the danger is in malfunctioning or misconstrued machinery or in an emerging inimical life-form; the difference is in what the ultimate *target* of our fear should be. In the first case, the fear is ultimately targeted on *ourselves*, focusing on the fact that we may at some point be unable to control our machines properly. In the second case, our fear goes beyond that: it is ultimately targeted at *them*, focusing on the possibility that *they* may turn against *us*.

An analogy may illustrate this difference in affective attitudes. Frozen computer screens and malfunctioning software often create frustration, and many users sometimes experience situations in which they vent their frustration with a fit of anger *at their computer* – perhaps pressing the Esc-key extra hard in retaliation, perhaps swearing a bit at the damned thing. Few, however, would go as far as to claim that any such rage against the machine is somehow *adequate to* or even *required by* the nature of the problem with which they are faced. Our anger should be directed at the manufacturers and programmers rather than at the machine – the machine itself is not a suitable target for anger, because it is not the sort of entity towards which anger makes any sense. Anger tracks wrongdoing – the machine, however, does not do anything wrong, or right, in the sense of the word in which wrongdoing is different from malfunctioning. And this is the difference emphasized by the skeptical "robophilosophy" that originated in the 1950s emphasizes: machines function or don't function, but they are not susceptible to normative reasons in the way required for participation in discursive practices.

Against this background, the kind of fear of robots that is not just targeted on what we might end up doing in putting robots to work, but on future robots *themselves*, seems irrational in the same way as an ordinary fit of anger against our computer. From now on, let's call the first kind "fear of robo-technology", and reserve the term "robophobia" for the second kind. Robophobia is the kind of fear of robots where the ultimate target is *them*, the robots. These two fears have different targets, and it seems obvious that choosing the right target does matter – though perhaps not in the finale of some dystopian scenario. Capek's play invokes reasons to assume a slippery slope from fear of robo-technology to robophobia over possible future courses of robo-technology. By contrast, robo-skeptics suggest that robophobia is inadequate on conceptual or quasi-metaphysical grounds, an irrational attitude that distracts from the real issues.

In the following, I will first elaborate on some aspects of the first skeptical claim, especially the claim that robots borrow the *form* of their actions from *us*, that is, the society in which they operate, and are thus dependent in their very agency on *our* life-form. Looking at the development robot technology is currently taking, I will argue that this view has lost a great deal of intuitive plausibility. It becomes increasingly likely that in some selected domains at least, robots may become able to fully participate in, and actively contribute to, the kind of discursive or proto-discursive social practices that provide the form of action – at least if these practices are understood in the way advanced in some of the literature. The (admittedly rather limited) case of robot cars do seem to make such a scenario quite realistic, and once robots are recognized as potential participants (rather than dependent appliances) in *our* life-form, it is hard to see from this perspective why it should be impossible *in principle* that robots should not be able to end up with a life-form of their own. This challenges or even undercuts the conceptual distinction between "mechanical" or "standardized work" and "discursive practice" on which so much of earlier skeptical "proto-robophilosophy" seems to rely (Sect. 12.2).

Yet not only robot technology, but philosophical research, too, has advanced over the last couple of decades. In the second section, I explore the possibility of supplementing and strengthening the skeptical argument concerning the life-form dependency of action with the old claim of life-dependency of life-form, focusing on the relation between *mechanisms* and *organisms*. I argue that the view that participants in life-forms need to be *living* creatures is sound, and I briefly invoke

Michael Thompson's 2008 recent analysis of how agency is based on life in this section. It is not obvious, however, that this should rule out robot life-forms a priori – here, a challenge is to be taken on from Capek's robots. They are *mechanical artifacts and living creatures at the same time*, thus undercutting another conceptual distinction which in potential future developments may prove to be metaphysically less robust than it appears (Sect. 12.3). The emerging view in these two sections is that as a "thick" concept, it is the history of the concept that shapes its content, and that the ambitions of some received "proto-robophilosophy" to radically re-evaluate the concept may stand on rather weak grounds.

The concluding section of this paper focuses on the eerie final scene in the epilogue of Capek's play and argues for seeing it as an intuition pump for the core normative issue that seems to be part and parcel of the very concept of "robot", that is, the nature of the displacement of humans by robots (Sect. 12.4).

12.2 Robot Routines vs. Discursive Practices

In order to see what kind of creatures Capek's robots are, it is important to observe them closely and see what they do. The introductory scene of Capek's play shows the owner of the robot-producing company, a man by the telling name Harry Domin, in his headquarters, where he interacts with one of his creatures, Sulla. Domin dictates a letter to a business partner, Sulla acts as his secretary writing it down. Sulla – a "female robot", meaning that robots make their first appearance in our culture in a gendered form – is obviously the one in the know about the tasks ahead, and Domin, having finished the dictating, addresses her. "Is that it now?", Domin asks her; "yes", Sulla replies, before another robot – Marius – enters the scene, seeming to act as a sort of impromptu butler: "Mr. Domin, there is a lady outside who is asking... ". Domin interrupts: "Who is it?". "I do not know", Marius replies, handing Domin the visitor's card (is he competently following a rule of etiquette by not nosing about his Master's visitor's identity, or has he simply forgotten to read the card?).

Capek's robots are very much like the real robots of our times in some respects, but very different indeed in others. Wikipedia informs us that the Czech word for "forced work" or "slave labor" is at the linguistic core of Capek's neologism, and the idea that robots are produced to do work (for us) is, of course, still at the very heart of our conception of robots. In this respect, Capek's fiction has become part and parcel of our economic reality, and certainly with increasing importance in our future. Our assembly lines have evolved from Fordean (or Taylorean) standardized human labor through Volvo-style flexible team work (as introduced and widely copied in the 1970s) to a heavily robotized machinery that can adapt to changes in the models to be produced. "Automatization of industrial production" is the key term here, and this development is certainly not unimportant for what we have come to mean by "robot". According to ordinary language, automata are mechanical devices, and they are the kind of mechanical devices that have an increased degree of independence (such as in the paradigm case of an automatic mechanical watch that you don't need to wind up yourself). Not every automaton is a robot, though the distinction between robots and other automata seems to be somewhat fuzzy. Robots are automata that perform *complex tasks* (or the automatic equivalent thereof, if you resist the idea that an automaton can properly "perform a task"), in something of the fashion of a human worker. It is notoriously difficult to specify what exactly this means, but a reasonable way to put this seems to be that robots need to be able to perform complex tasks in sufficiently different ways or to perform *a range* of tasks rather than just one. Robots need not – and perhaps cannot – be "universal", but if there is only *one thing* an automaton does, it cannot be a robot.

Industrial robots in the assembly lines of production can serve as a paradigm here. These robots are basically mechanisms developed to replace human workers in standardized tasks, and they are capable of coping, to some degree at least, with changes in the assignment of their task, so that the same robots can produce different models. Robotization has significantly changed the sense in which our societies might be called "post-industrial" (Bell, 1974); it seems to be due to robots that not all industrial production has gone to developing countries, where human labor is cheaper (the robots would certainly have deserved a bit more than being mentioned just once and in passing in the new foreword to the 1999 edition of Bell's study). Theories of post-industrial society are heavily based on the three-sector theory (e.g., Fisher, 1939), which distinguishes the domains of economic activity in the extraction of raw materials (farming, mining etc.), manufacturing (production and construction), and the service sector (providing what is sometimes called "intangible goods", where knowledge, advice, and attention are at the core of what is bought and sold). The three-sector theory is well established not only in economics, but has come to be widely known and accepted in the general public. It is basically a developmental theory; the view it entails is a distinction between three *kinds* of work and the claim that over the course of economic development, the main focus of economic activity passes from the primary through the secondary to the tertiary sector.

A good way to approach the question of what robots can and cannot do is to use the three-sector distinction of kinds of activity: extraction, production, service. Besides the very conspicuous robots for the secondary sector, there are already robots developed for agriculture (harvesting robots, driverless tractors etc.) and mining (deep sea robots seem to be ready to exploit the resources to be found there), but neither of these sectors is where we first meet robots in Capek's play. Classic industrial production is not the kind of work Marius and Sulla are doing in Capek's opening scene, even though Capek suggests that industrial production at the time of his play is largely in the hands of robots, and in spite of the fact that in the epilogue, it is suggested that the robots identify their own life-form as one of production and manual work. The last human is spared in part because he is seen, by the robots, as robot-like: "He is a worker. He works with his hands like a robot. He builds houses. He can work." Marius and Sulla, however, are not engaged in production at all. They are doing *service work*, even though they offer their services within a production plant. And this raises questions concerning their identity: if they are successful at such tasks, can they *really* be robots?

Sulla not only types the letter, but serves Domin with her knowledge of the tasks ahead and her capacity to tell what's urgent from what can wait in the given situation. She acts context-sensitively with the very kind of "knowledge of what matters in the situation" that is often denied to robots in the received literature. A *locus classicus* for this is Daniel Dennett's (1984) robot who is programmed to tell what's relevant from what's irrelevant, but can do so only by going through its entire knowledge base and is thus always too late to act, lacking the capacity for *emotion* that provides human agents with a quick and dirty estimate of relevance. Sulla seems to have no problem there. And similarly for Marius' first appearance: An important aspect of what makes his presence valuable is that he serves as a trustworthy, well-mannered and agreeable butler who does not stick to some scripted protocol, but interprets his role competently. He interrupts his announcement when Domin asks him who the visitor is, he knows that handing over the business card rather than reading it himself in Domin's presence is the appropriate thing to do. Marius seems to know that "letting Domin know who the visitor is" can be done in many different ways and that if the preferred way is blocked, there are alternatives, and that there are standards of propriety that guide the right choice of alternative in a given context with a sense of which rules matter and which do not. Can *this* be programmed, and thus mechanized, or is it rather part and parcel of the very idea of programming and mechanization that behavior is standardized and thus not the kind of behavior in question? The established way of spelling out the skeptical view draws on Michael Polanyi's *Personal Knowledge* (1958). Polanyi presents his skepticism in arguing – against Alan Turing – that robots may *simulate* such features as mindedness and personhood, but that any such appearance will always be *deceptive* (1958, 277). Whatever actions robots may simulate, they are simulations of actions and not the real deal. This may not matter in *some* domains of action and seems irrelevant in domains of activity where the focus of interest is on the results, such as in extraction and production, but it does matter in domains of activity where the focus extends to the *form* of activity, such as in service. Polanyi argues that real agents operate on a *different kind of knowledge* than their mechanical counterparts. The key terms here are *inarticulate*, *personal*, or *tacit knowledge* – other labels are *implicit* or *social knowledge*. And this sort of knowledge is decisive where work is not *just* a means to an end, but also involves partnership in interaction.

Polanyi's view has been developed further by Harry Collins in his *Artificial Experts* (1990). The point here is not that behavior involving implicit knowledge cannot be mechanized. Indeed, Polanyi's paradigm for a practice based on implicit knowledge, riding a bicycle, will hardly be much of a problem even for future robots. Rather, the point is that any such behavior, however similar it may seem to its non-mechanical counterparts, will be of a profoundly different kind. In Harry Collins' and Martin Kusch's *Shape of Action* (1995), this is spelled out in the distinction between *mimeographic* and *polymorphic* action: Mimeomorphic action may involve a great deal of tacit knowledge, but it can be mechanized; polymorphic action, by contrast, requires knowledge that is not only implicit, but also rapidly evolving and situationally flexible with the shared practices of a group. *Social knowledge* of the kind at work in polymorphic action is a matter of life-form and cannot be "fed into a machine" in an algorithmic fashion, because it does not reside in individuals at all, but rather in groups – it has a holistic structure that cannot be divided into independent components and fed into computers. It involves, as Polanyi put it, "conviviality", "a cultural life shared by a community", social exchange and social assent (1958, Chap. 7) – in other words, *our* life-form. An example of this distinction is between Polanyi's case of simply riding a bicycle in isolation – a mimeomorphic action – and riding a bicycle through heavy traffic, where unexpected situations often occur and where ways of dealing with them have to be negotiated between the participants – a polymorphic case (Collins & Kusch, 1995, 87).

Collins' and Kusch's distinction is helpful in that it adds an important distinction into received taxonomies of kinds of action. Yet it is not entirely obvious how well it is suited for Collins' robo-skeptical agenda. Let us focus on the paradigm example of public traffic. Our days have seen the first robot cars, as several automobile manufacturers are testing prototypes, which seem to be quite successful and are announced to go into production soon. If, in all likelihood, no robot bicyclist will be seen on our streets in the foreseeable future, this seems to be less a matter of the kind of action at stake, but of the lack of purpose for such a robot – not impossible at all, but simply pointless. There is no reason to assume that robot bicyclists would be less successful in moving through heavy inner city traffic as robot cars already are.

From a skeptical point of view, this may just show that moving through traffic can be *mechanized* and thus *mimeomorphized*, as it were, and that the domain of one action form may be extended over domains which were previously reserved for action of the polymorphic kind. Yet the skeptical distinction between a mechanic *simulation* of the behavior of a competent participant in the social practice of moving through traffic and *real* participation in the social practice of moving through traffic seems to collapse as soon as robots not only actively use available information from central sources and continually update their algorithms, so that their behavior is in permanent change, but also give feedback and *negotiate* ways in which unusual situations are to be solved on a case-basis. The relation between robot routine and discursive practice may turn out not be of the disjunctive kind after all. Imagine the following scenario: A (future) robot car registers a pedestrian moving towards the curbstone, with no pedestrian passing in front of him. Assuming that the pedestrian wants to pass, the car slows down. As the pedestrian stops at the curbstone, the car gives a signal with its headlights. The pedestrian makes a waving sign to the car to pass by. The car registers this and moves on. It does not seem obvious that a future car may not be able to *learn* to interpret such signs; moreover, it does not seem all too far-fetched to assume that in our interaction with traffic robots, and especially in their interaction between themselves, new ways of coordination and practices of getting by each other will be negotiated. It is not unlikely that the discursive practice of participation in public traffic, the forms of getting along and getting by will evolve further based on robot contributions to the practice. Such robots, it seems, are not simply "mechanical simulations" of participants in a social practice in which polymorphic action is mimeomorphized. They are themselves active participants in the maintenance and further development of this practice. At least as far as the special case of traffic is concerned, they are active participants of a "cultural life shared in a community" – why, then, should they not be seen as *members* of the society who is the locus of the relevant implicit knowledge, which they so actively co-constitute?

In the light of recent developments and the speculations about possible futures they seem to support, the case of the social practice of coping with traffic may not seem particularly apt to support the skeptical view. The hierarchy between "mimeomorphic robots" and "polymorphic human community members" may change; *they* may come to cope with traffic problems in much more flexible and dynamically negotiated ways, leaving us the role of participating in *their* culture in a way that may appear as a mere stereotypized *simulation* of what they are doing so competently. The distinction between the mimeographic and the polymorphic may turn out to be a question of social status, and within the practice of negotiating ways of getting by each other in public traffic, it may soon emerge that robots are perceived as being "better at the task", leaving us the role of the mimeomorphic simulators of our robot interaction partners.

It is a very far shot from here to the dystopian scenario within which the term "robot" was introduced, but the case seems to poke some holes in the barrier that roboskepticism has erected. And perhaps the roboskeptical dichotomy is the *problem* negotiated in Capek's play, rather than the solution. Perhaps the distinction between "mere robot routines" and "proper human discursive practices" involves a problematic status system. Let us speculate a bit further along these lines. "Domin" is not only Marius' and Sulla's manufacturer who uses them as particularly complex mechanical tools. He is in fact participating, together with them, in a *shared social practice* of managing his office – a social practice in which he, as his name suggests, assumes the role of the ruler, and it is perhaps because he ignores that *their* actions co-constitute the shape of his own actions and disrespects the "conviviality" (Polanyi) he shares with his servant creatures, that he – as well as his human conspecifics – miss the opportunity to negotiate, together with them, an arrangement that is agreeable to all participants. At the end of the play, Capek's robots rise in rebellion, starting to shape their actions in *their own way*.

This may not seem intuitively plausible – but how not to end up there as a *possible* scenario, once the distinction between robot routine and discursive practice has started to collapse? The conclusion seems to be that the conceptual "line of defense" raised against any such scenario by skeptical robophilosophy may not stand as firm as it may appear.

12.3 Mechanical Artifacts vs. Living Organisms

That robots are rarely seen until now in our service economy is likely to change very soon. The virtual relatives of robots, the software-bots, seem to be developing at an impressive speed. Admittedly, the most advanced chatbot experience that is currently (2016) advertised on the internet seems to be rather far off the Turing test mark, but at least occasionally, a surprising few lines of exchange are already possible with the likes of Mitsuku[3] (while infamously, Microsoft's Tay has proven to be all too human, in a way, by developing into something of an openly racist holocaust denier within a few days of its short existence). It seems that within tightly limited domains of discourse, some bots have already passed for humans (the Turing test entails no such restrictions). It is likely that soon enough, you won't know if you're talking to a human or a program if you call a customer service. And it does not seem far-fetched to assume that physical robots will soon enter our service economy, too. PARO, the therapeutic robot seal, already improves the elderly patients' mood and their interaction with other patients and caregivers, and although PARO's role is similar to that of a therapeutic cat or dog and nowhere near the kind of work a nurse or human caregiver provides, it is to be expected that caregiving will be among the first areas of service economy that will be increasingly robotized with physical robots. Such robots might play human service work roles somewhere along the lines of Capek's Marius and Sulla, even though their first appearance might be in retirement homes rather than CEOs' offices.

As physical robots are likely to graduate from manufacturing to service very soon, not only their capabilities, but their physical appearance, too, is likely to change. A decisive difference between robots working in industrial production and future robots working in the service economy is that the latter will have to play the role of interaction *partners*. And even though this may not always be the case, it is to be expected that for this purpose, they will have to be human-like in the relevant domain of interaction (e.g., Duffy, 2003).

This is obviously already true for Capek's robots. In the stage directions, Capek introduces the robots as "slightly mechanical in their speech and movements" and as "blank of expression, fixed in their gaze", but over the course of the play, they come to express a great deal of thought, they basically act like ordinary humans, finally including the full repertoire of emotions; they behave very much like the human characters, and in one scene of the second act, a human mistakes a group of humans for robots. Even though only few existent robots come anywhere close to any such human likeness, it is certainly something many think our robots could be able to do, in some form or another, in the not-so-far future (Hiroshi Ishiguro's Geminoids already seem surprisingly real, at least on youtube-videos).

Yet Polanyi's skeptical intuition that whatever actions a robot might eventually come to be able to perform, it will have to *deceive* us in order to pass as "one of us", will certainly not be off the table. Fear of simulation is not alleviated by similarity – quite to the contrary: the likelier, the scarier, and the more vivid the fear of being *deceived*. One way of spelling out this attitude is along the following lines: The kinds of things we do are part of our life-form, and a minimal (yet not sufficient) requirement to participate fully in our life-form is just that: to be alive. Robots, however, are not alive. They are not organisms, but mechanisms. If robots are a

[3]http://www.mitsuku.com/, encountered on 8/19/2016.

kind of automata and if automata are a kind of mechanical devices, how could they possibly be *alive*?

In an influential article from 1959, Paul Ziff argued that whatever is said about a robot's reasoning, thinking, feeling, and acting "means whatever it usually means except that one cannot infer that what is in question is a living creature" (Ziff, 1959, 67). For Ziff, the distinction between organism and mechanism is fundamental, and he illustrates this with the distinction between how an organism dissolves in its environment after its death, and how a mechanism is taken apart or falls apart. Ziff suggests that the organic and the mechanical are mutually exclusive, and that it is somehow due to the fact that robots are mechanisms rather than organisms that "there are no psychological truths about robots" (ibid.), and that they cannot act.

In the general discourse, the terms "organic" (and thus living) and "mechanical" seem to operate as mutually exclusive terms, so that if something is an organism, it is thereby not a mechanism, and vice versa. In this way, this distinction has often been invoked as decisive in the skeptical camp of the debate on robotics and artificial intelligence, at least since the 1950s. But how sound is this premise? In the first couple of sentences of the introduction of his *Leviathan* (1651), Thomas Hobbes famously claimed that "life is but a motion of limbs", reducing life to *mechanics* and thus undercutting the usual disjunction right off the gate. He continues by asking the provocative rhetorical question: "why may we not say that all automata (engines that move themselves by springs and wheels as doth a watch) have a life? For what is the heart but a spring; and the nerves, but so many strings; and the joints, but so many wheels, giving motion to the whole body, such as was intended by the Artificer?" Chapter xiii of the *Leviathan* takes this mechanical analysis from the anatomy to psychology, giving a mechanical analysis of psychological features such as the human emotions. And whereas Ziff takes the difference between an organism's *dying* and a mechanism's *falling apart* to be obvious, there is none in Hobbes' view; an organism's dying is nothing but a mechanism's falling apart, according to Hobbes.

Hobbes thus asserts that the "art of man [...] can make an artificial animal" – the very point of which Ziff seems to assume being impossible. Why should we accept Ziff's premise rather than Hobbes'? Why not say, against Ziff, that the big difference between our relation to organisms and artificial automata expresses no more than a sort of parochialism about the value of different kinds of automata?

That there are different *kinds* of automata in question here, however, is not ignored by Hobbes, and indeed he places a great deal of weight on the distinction between "natural" and "artificial" life. "Natural" life, Hobbes argues at the beginning of the *Leviathan*, is different from the "artificial" kind in that in the first case, the "Artificer" is God, in the second case, humans, and it is very important to see the *kind* of "artificial life" on which the *Leviathan* focuses. Except for the very beginning of the *Leviathan*, Hobbes is not concerned with robot-like automata; rather, he is concerned with social *organization*. Robots are artificial agents, but they are artificial agents of a very different kind than the political bodies Hobbes has in mind under that title. They consist of mechanical parts rather than natural persons, and they include machines that are not rooted in human consensus and shared interests in the way Hobbes' "personae fictae" are. In spite of his introduction, Hobbes is not concerned with robots at all. The fact that he ascribes life in quite a literal sense to *social organizations* and places all his hopes for mankind's future on them, certainly does not mean that he had been a robo-enthusiast, had the concept ever crossed his mind. Whereas Hobbes' whole point seems to be that our "natural" life, for its preservation, requires the creation of artificial life (of the kind of human organization), because natural organisms such as us cannot *coexist* without creating artificial life, no such a priori link exists concerning the artificial life of robots. As Hobbes' frontispiece illustrates quite vividly, his "artificial agents" are *made of us*, and the life of the Leviathan is an artificial (yet rationally required) form of *our* life. This obviously does not carry over to artificial agents such as robots in the same way, and there is thus no way of telling how Hobbes would have reacted to the challenges posed by our emerging robo-technology. He might well not be using the analogy between humans and automata in the same way anymore and spell out the *differences* between "natural" agents, and social organizations, and robots. If we apply the term "life" to all of them, it is obviously in a very different sense that they are "living", and the relation between "our" natural lives and the artificial lives of organizations and robots – as well as the relation between social organizations and robots – is problematic in very different ways indeed.

A Hobbesian-minded way to spell out the difference between natural life and the different kinds of artificial life focuses on the stuff of which these agents are made – which is the flesh and blood of natural agents, natural agents of Hobbesian organizations, and steel, silicone, and rubber in the case of robots. But how fair is it to strike an entity from the list of living creatures, and thus proper agents, simply because of the material it happens to be made of? Doing so seems to express the worst kind of focusing on what an agent *is* rather than taking into account what he *does*. Isn't this a very parochial attitude indeed? Chemists sometimes claim that life could have evolved on the base of silicone rather than carbon – why should we dismiss any such view from the outset?

A forceful voice in the recent philosophical debate on life is Michael Thompson's book by the title *Life and Action* (2008). Thompson is neither concerned with Hobbesian reductionism nor with robo-technology here, but his perspective might be interesting for the kind of robo-skepticism initiated by Ziff in that he makes a forceful argument for accepting "life" as *irreducible*. The Hobbesian "motion of limbs" is obviously insufficient as a reduction base for life of the *organic* kind because life involves the motion of limbs only insofar as the limbs in question are *an organism's*. And similarly with regard to the candidates often invoked by reductionists in the more recent discourse. Life cannot be reduced to such features as homeostasis and reproduction, because whatever defining feature of life is identified can be found in non-living instances (e.g., homeostasis in corpses, reproduction in crystals). If life is a matter of maintenance or reproduction, Thompson argues, it is a matter of the maintenance or reproduction of something of the *form of life*. "Form" is the central term in Thompson's reductionism, and Thompson approaches the question of what a form is from the perspective of the corresponding judgment. The question of what a life-form is, is thus approached in Thompson's book by looking at "how these are described", and this is done in what Thompson calls "natural-historical judgments" (62). These are judgments of the sort "the cat has four legs". Judgments of this form, Thompson argues, are "a logically special form of appearance of predicative expression" (76), and whatever noun is "suited to be the subject of such predication" is thereby a *"life-form word"*.

The label Thompson chooses – natural-historical judgment – seems to suggest that we need to imply some history in order to describe something as living, which would rule out robot life on conceptual ground, as such life is *created.* As far as I can see, however, Thompson does not endorse any such neophobic prejudice per se. Rather, he argues that life-form judgments involve a unique relation between species and exemplar involved in judgments of the form in question. "Rocks are hard" applies to *this* nugget in a different way than "cats have four legs" applies to *this* cat, and this comes to the fore in the way the species can be meant in such statements as "when springtime comes, and the snow begins to melt, the female bobcat gives birth to two to four cubs". Does that rule out robot life, because there would be a single first living robot, without any species to which it belongs? A Thompsonian-minded answer might be that the species is implied in the judgment, and would be there *potentialiter*, as it were. What would be lacking in this case would be an important dimension of life-form judgments: their proto-normative nature. The way species and exemplar are related in the form of natural-historical judgments involves a *standard* against which exemplars are held. Of *singular* creatures – take Stanislaw Lem's living planet Solaris as an example – we cannot grasp their form as of the kind of life unless we assume something about what living planets are, in general – "normally", as it were.

The question thus is: Is "robot" a life-form-word? Where is the logical or formal difference between "when springtime comes, and the snow begins to melt, the female bobcat gives birth to two to four cubs" and such judgments as "when daylight is gone and the lights are switched off in its nursery home, PARO relaxes and closes its eyes to sleep"?

Obviously, PARO has no natural history of its own. PARO is a recent invention, and whatever natural history plays a role in the way we describe PARO is the original seal's. But the logical form Thompson has in mind explains whatever the natural history in the label of the respective judgments may mean, and not the other way around. And here, Thompson focuses, on the one hand, on the relation between exemplar and species, and, on the other hand, the sense in which natural-historical judgments are proto-normative, that is, involve a conception of the *ordinary* or *normal.* But in our judgment about PARO we seem to take *this* individual as exemplar of its species in the same way as the discovery channel zooms in on *this* bobcat in explaining what bobcats do. And it seems that if PARO does not close its eyes in darkness etc., a deviation from the standard of the ordinary or normal happens in the same way as in the case of the bobcat that does not breed.

Thompson is concerned with the form of natural-historical judgments, and not with the demarcational question at stake here, that is the question of whether and under what conditions robots may be thus judged. One point to make for the exclusion of robots may perhaps be that in their case, the standard of normality to which robots are held is set by *us*, and it is *up to us* in a way that is not the case in "natural" organic life. Perhaps all judgments concerning our artifacts share the Thompsonean form-features mentioned above – take the case of a musical instrument as an example. But there, as in the case of any functioning of our machinery, the standard is entirely up to us. It is in the light of our attitudes that the standard holds. This is not the case in organic life; the respective proto-norm is ultimately a feature of the target of our judgments in a way it is not in the other cases. If closing its eyes in darkness is an accepted feature of PARO's, that's that about the norm; the acceptance *constitutes* the norm in a way that has no equivalent in organic life, where we may *think* of a life-form to have a standard which it does not in fact have, and where the standard is external to our attitude. In this way, the point would, again, be that whatever life-form there is in judgments about robots is *derivative* from our life-form, so that the idea of a robot life-form *displacing* ours appears, again, as inconsistent. This view, however, is challenged by another feature of Capek's robots. They are mechanisms and organisms *at the same time*, and the question raised by this feature is: Why should something that is defined as what it is with reference to an imposed function not have a function *of its own*, at the same time? Why may robots, whose standard of what's normal about them is set by us, not develop into a proper life-form with standards of its own?

The idea that the conceptual distinction between *mechanism* and *organism*, between artifact and living being may not be clear-cut at all is built into the very conception of the robot, if we take it to be determined by its history. Capek's robots involve, from the very beginning, an element of *life*. Capek's robots are, in some sense of the word, *living organisms*. They are made of "synthetic protoplasm", and Capek has his Domin say that the robot's tissues behave just as ours, "despite being, chemically, quite different": "Nature has found only one way of organizing living matter. There is however another way which is simpler, easier to mould, and quicker to produce than Nature ever stumbled across." It seems that in some ways, Capek's robots are closer to such conceptions as the *replicants* in Ridley Scott's 1982 movie *Blade Runner*, which are organic and produced by genetic engineering, but Domin's remark suggests that the physical substrate of the robot's organic life is different from nature's, and "moulding" rather than genetic engineering is quoted as the way of their production. In this sense, the way in which Capek conceives of "robots" crosses the distinction between the mechanical and the organic, and in spite of first attempts to construe "biots" – mechanisms that include biological parts such as muscle fibers – the idea of *producing* creatures that are *entirely* made of biological parts has come to fall in quite another domain, the practice of genetic engineering, with problems and questions attached to it that we have come to distinguish from the problems and questions of robotization. At a time in which genetic engineering has advanced, the conceptual distinction between "mechanism" and "artificial organisms" may come down to the distinction between two different sorts of technology – but how confident should we be that these branches of technology will never converge in the future? Capek's robots challenge the conceptual dichotomies that have been drawn by robo-sceptics. They undercut the distinction between robot routine and human discursive practice as well as the distinction between mechanism and organism. And perhaps this very challenge is at the core of the idea of a robot.

12.4 The Capekian Scenario as an Intuition Pump

In the epilogue of Capek's play, the last surviving human being, the robot's constructor Almquist, witnesses the last transformation of his robots, and the robophobic sentiment that developed over the three acts of the play is suddenly lifted. A male and a female robot specimen, Primus and Helena, develop a sense that "they belong to each other", as they put it, and to Almquist, who is about to dissect one of them in order to retrieve the formula for their production which he had lost earlier in the play, they prove to be willing to sacrifice their lives for each other, each one begging to be the one to be killed. Almquist recognizes a new Adam and Eve in his robots, he lets them go and, invoking the sixth day of creation, he sends them out to populate the earth in place of the all but distinct human life-form. With this, Capek's robophobic play ends with an eerie sense of hope. The scene may seem a bit crass, and to many it may seem absurd rather than touching, but the question of the success condition of this last plot twist drives the central point in the whole idea of a robot life-form home rather nicely: The concluding scene suggests conditions under which robots can be properly judged to be a life-form *of their own*, independently of the question of their servile or hostile relations to mankind that was negotiated in the previous acts. The epilogue presents robot existence as developing into something that can be seen as having *intrinsic value* (and particularly in form of a thriving shared social life), and for which there can be *hope*. The scene suggests that now the robot life-form has become something that *should* be, and whose function is wholly intrinsic, rather than dependent on a value that determines its function *for humans*. What do we have to assume of the robots in question for *this* attitude to be adequate?

I submit that Capek's scenario is a good way to approach this central question of robophilosophy, and perhaps it would be a good thing to pursue it with the instruments of experimental philosophy. Slight variations of Capek's final scenario might change people's view, and it would be most interesting to see what – if anything – people would see as a *reason* for hope for the robot life-form. Capek suggests a clear answer to this question, and his may not be the most convincing one. At this point, the play alludes to the classic values of *romantic love*. A somewhat Aristotelian interpretation may be more plausible here: In order to be an independent life-form of their own, with fully intrinsic value, the robots need to be striving towards the ultimate good of a thriving communal life. Yet it seems that *any* life, however low it may be, has some intrinsic value (there are many reasons to kill organisms, but if there is no such reason, the fact that an organism is alive is a reason not to kill it). Why can't the unromantic robots be recognized as an intrinsically valuable life-form, sent out to populate the world in their previous frame of mind, without recognizing intrinsic value in their togetherness? The interpretation that suggests itself is that without the ultimate romantic turn, robot life would not be *our* life, or a life-form that is structured by our values. The ultimate scene – that is repeated in so many other novels and plays that revolve around the issue – seems to play with the ambivalence it arouses in the audience. It finally gives the strongest reason to accept a robot life-form, the capacity for romantic love, yet at the same time it somehow raises the robophobic horror to its culmination: There is always the intuition that this last turn only completes the tragic human delusion, total human self-alienation in a fetishized view of their products. Almquist dies sensing that "life will not perish", and there is as much to say for it as against it – Capek's drama, as so many later stories that followed this plot, obviously plays with this ambivalence.

Capek's final scenario may serve as an intuition pump in that it may help to identify conditions under which hope or utter despair is the adequate attitude. Contrast Capek's concluding scene with the following scenario: Human life goes extinct because the fertility rate declines. The last human beings spend the rest of their lives in old age homes, being cared for by robots of the sort of PARO, only that they now include further care features. You are the last human in your old age home, and you witness how those future PAROs turn from caring for human individuals to caring for their damaged conspecifics and start to engage heavily in PARO-manufacturing, so that there is reason to assume that the world will soon be crowded with PAROs. It seems that whatever plausibility Almquist's attitude towards Primus and Helena has is certainly lost on this strange PARO-community: You would hardly die with a sense that "life will not perish", as Almquist does. Your feeling would probably rather be that a mechanism set off by humans, and which has gotten out of hand, will simply continue to run as a perpetuum mobile of sorts, but there is no sense in which there is any good to it, except perhaps that the surviving artifacts may bear some form of witness to the extinct human life-form. In the PARO case, any value or disvalue there is, is derivative of human life. What has to be added to PARO and its conspecifics for you to feel differently? In Capek's case, it is suggested that robot life would basically have to be of *our* form, but this seems parochial and speciecist – which opens the debate.

Capek's core point does not seem to be that it gives one (Almquist's) or the other (the horrified audience's) answer. The point is in the ambivalence. And I submit that it is in *this* sense that "robot", in the Capekian tradition, is a life-form word. It is a life-form word in that it confronts us with the messiness of the basic conceptual distinctions of human self-understanding in an increasingly engineered world, the collapse of dichotomies such as "routinized behavior" vs. "discursive practice", or "artificial mechanism" vs. "organic life" is the message here. And in contrast to Capek's play, we can't just wait and see what happens, in order to find out which attitude we may find adequate.

Acknowledgements I am grateful to Johanna Seibt and Raul Hakli for many insightful comments.

References

Bell, D. (1974). *The coming of post-industrial society: A venture in social forecasting*. New York: Basic Books.
Collins, H. M. (1990). *Artificial experts: Social knowledge and intelligent machines*. Cambridge, MA: MIT Press.
Collins, H. M., & Kusch, M. (1995). Two kinds of action: A phenomenological study. *Philosophy and Phenomenological Research*, *55*, 799–819.
Dennett, D. C. (1984). Cognitive wheels: The frame problem in AI. In C. Hookway (Ed.), *Minds, machines, and evolution: Philosophical studies* (pp. 129–152). Cambridge: Cambridge University Press.
Duffy, B. R. (2003). Anthropomorphism and the social robot. *Robotics and Autonomous Systems*, *42*, 170–190.
Fisher, A. (1939). Production, primary, secondary and tertiary. *Economic Record*, *15*(1), 24–38.
Hobbes, T. (1969). *Leviathan, 1651*. Menston: Scolar Press.
Polanyi, M. (1958). *Personal knowledge: Towards a post-critical philosophy*. London: Routledge.
Seibt, J., Hakli, R., & Nørskov, M. (Eds.). (2014). *Sociable robots and the future of social relations: Proceedings of robo-philosophy 2014* (Vol. 273). Amsterdam: IOS Press.
Thompson, M. (2008). *Life and action*. Cambridge, MA: Harvard University Press.
Ziff, P. (1959). The feelings of robots. *Analysis*, *19*, 64–69.

Author Index

R. Hakli, J. Seibt (eds.), *Sociality and Normativity for Robots*, Studies in the Philosophy of Sociality 9, DOI 10.1007/978-3-319-53133-5

S

T

U

V

Subject Index

R. Hakli, J. Seibt (eds.), *Sociality and Normativity for Robots*, Studies in the Philosophy of Sociality 9, DOI 10.1007/978-3-319-53133-5

J

K

V

W

MIX
Papier aus verantwortungsvollen Quellen
Paper from responsible sources
FSC® C105338

If you have any concerns about our products,
you can contact us on
ProductSafety@springernature.com

In case Publisher is established outside the EU,
the EU authorized representative is:
Springer Nature Customer Service Center GmbH
Europaplatz 3, 69115 Heidelberg, Germany

Printed by Libri Plureos GmbH
in Hamburg, Germany